P9-EEG-495

Testing UMTS

Testing UMTS

Assuring Conformance and Quality of UMTS User Equipment

Dan Fox

Anritsu

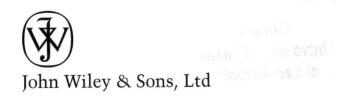

John Wiley & Sons, Ltd

Other Wiley Editorial Offices

John Wiley & Sons Inc., 111 River Street, Hoboken, NJ 07030, USA

Jossey-Bass, 989 Market Street, San Francisco, CA 94103-1741, USA

Wiley-VCH Verlag GmbH, Boschstr. 12, D-69469 Weinheim, Germany

John Wiley & Sons Australia Ltd, 42 McDougall Street, Milton, Queensland 4064, Australia

John Wiley & Sons (Asia) Pte Ltd, 2 Clementi Loop #02-01, Jin Xing Distripark, Singapore 129809

John Wiley & Sons Canada Ltd, 6045 Freemont Blvd, Mississauga, ONT, L5R 4J3, Canada

Wiley also publishes its books in a variety of electronic formats. Some content that appears in print may not be
available in electronic books.

British Library Cataloguing in Publication Data

A catalogue record for this book is available from the British Library

ISBN 978-0-470-72442-2 (H/B)

Typeset in 10/12pt Times by Integra Software Services Pvt. Ltd, Pondicherry, India.
Printed and bound in Great Britain by Antony Rowe Ltd, Chippenham, England.
This book is printed on acid-free paper.

Contents

Preface

A modern mobile phone is a highly complex electronic system made up from a variety of diverse subsystems, all of which must work seamlessly together. Today's users have very high expectations set through years of experience using mature second-generation phones, and this sets tough demands on manufacturers as they introduce third-generation technology. While quality, in terms of the phone's stability, performance and behaviour on the network originate from good design, the only way to be sure of quality is by testing it. This makes testing a very important part of any mobile phone's life cycle, from development through to manufacture and beyond, touching a number of different disciplines and departments.

This book is intended to be of use to engineers and managers who are either directly involved in the process of testing Universal Mobile Telecommunications System (UMTS) mobiles, or who are looking for an understanding of what is involved in testing. In general, this applies to those involved in:

- The development of UMTS mobiles
- Integration and verification
- Conformance testing
- Operator acceptance testing
- Manufacturing
- Servicing.

The book is divided into three sections. Part I provides an overview of major types of testing and the organizations and tasks involved. In particular, it looks at what is involved in conformance testing and device certification. Part II is more technical and looks at the UMTS standard itself, working through the protocol layers. Where possible, I try to describe the operation of the standard from the perspective of testing it. Part III takes a brief look at some of the possible future trends and their impact on testing mobile devices, including the emergence of new technologies both in the access network and the core network and the evolution of new testing methodologies. The book concludes with an appendix containing a tutorial on Tree and Tabular Combined Notation version 2 (TTCN 2), a formal method of describing tests which is widely used in telecommunications protocol testing, and of particular relevance to UMTS.

Both UMTS and the testing of UMTS devices are huge areas. The 3rd-Generation Partnership Project (3GPP) standard consists of over 2200 specifications, and it is not possible to cover all of them in detail in this kind of book. I have tried to focus on the most important areas from a test perspective, but from necessity, there are many omissions. For full details, the reader will need to refer to the specifications.

Acknowledgements

I acknowledge the help and support by the staff at Anritsu who provided ideas and material for the book and reviewed the text. In particular, I thank Jonathan Borrill, Anritsu EMEA, for his assistance with information on manufacturing and service testing. I also thank the Anritsu product marketing team in Japan for help with much of the information used in the RF and system test sections and Paul Jeffs of Anritsu EMEA for providing various product images used to illustrate the text. I thank my main reviewers, Mike Lee and Ian Rose, of Anritsu UK for their assistance. Finally, I thank my family for their support and patience for putting up with all the lost evenings and weekends that writing a book demands.

Part I

Testing Methodologies

1

Introduction

In modern, complex telecommunications systems, quality is not something that can be added at the end of the development. Neither can quality be ensured just by design. Of course, designing for quality is very important, but no design process is good enough to guarantee that everything works correctly first time. The fact is that complex devices need extensive testing to make sure that they work reliably. Coupled with this, the limited selling window of consumer-oriented user equipment (UE) means that for high-quality products, this testing starts early in the development cycle and continues through to product deployment.

This gives rise to a wide variety of different tests and testing methodologies, applied at various stages in the development cycle. This book aims to provide an insight into achieving high-quality through testing, by providing the reader with both an appreciation of the various testing methodologies, and an understanding of how they fit together to make a complete approach to quality.

Modern networks are indeed very complex, and testing plays an important role in the development of all parts of the network. However, there are special considerations for the testing of the mobile terminals. These are deployed in large numbers; popular handsets can sell in the millions and are in the hands of users who are neither equipped for nor interested in their service and maintenance. This book is concerned mainly with the testing of these devices.

1.1 Important Definitions

The telecommunications industry, like many areas steeped in technology, tends to have a bewildering array of terminology specific to the industry. The Glossary covers the common acronyms and specialist technical terms used within the book, but there are a few key terms where a deeper explanation will assist understanding from the outset.

Testing UMTS: Assuring Conformance and Quality of UMTS User Equipment   Dan Fox
© 2008 John Wiley & Sons, Ltd

1.1.1 3rd-Generation Partnership Project

3GPP is the abbreviation for the 3rd-generation partnership project, an alliance of national and regional telecommunications standards organizations (the partners). The 3GPP is largely built on the foundations of the European Telecommunications Standards Institute (ETSI), based in Sophia Antipolis, France. ETSI takes care of administration and management of 3GPP, and the standardization very closely follows the principles and methodologies set by ETSI during the standardization of Group System Mobile (GSM) and other major European telecommunications standards.

While the term 3GPP strictly applies to the standardization body, it is also widely used to refer to the standard itself and is often used interchangeably with other terminology [e.g. Universal Mobile Telecommunications System (UMTS) or wideband code division multiple access (WCDMA)]. In this book, the term 3GPP will be used to refer to the standards body.

1.1.2 UMTS, UTRAN and GERAN

Conceptually, the UMTS originated as the European extension to the GSM system and was put forward under the IMT-2000 initiative of the International Telecommunications Union (ITU) as one of the converged family of standards for third-generation mobile communications. UMTS and 3GPP are often used interchangeably to refer to the standard, but in this book, UMTS will refer to the overall mobile telecommunications system as defined and standardized by the 3GPP. The UMTS system was designed to integrate with existing GSM networks. The network was split into two parts: a core network (CN) and a radio access network. The network architecture is described in more detail in Chapter 10. The CN from GSM was left essentially unchanged, and the standard defined a new radio access network to complement the existing one. This is known as the UMTS Radio Access Network, or UTRAN.

In parallel with the development of UMTS, the GSM network has also evolved to some extent. In part, it has evolved to provide more efficient use of the existing GSM spectrum, through the development of enhanced data rates for GSM evolution (EDGE) and Enhanced General Packet Radio System (EGPRS) which provide higher data rates and, in part, it has evolved in synergy with UMTS to allow operators with dual networks to offer new services more seamlessly. This evolved GSM radio access network is now referred to as GERAN–GSM/EDGE Radio Access Network.

1.1.3 User Equipment

UE is the official term for a device capable of interfacing to the UMTS network. In GSM, this was referred to as the mobile station (MS), and this term is still widely used in much of the documentation that is shared between GSM and UMTS. The term UE was selected mainly because UMTS was expected to include new classes of device beyond the mobile phones and hand-held computers of 2G technologies. It is intended to imply the broader span of devices that are expected to operate on a UMTS network.

1.2 Scope

Mobile communications is a very diverse field, with many different standards and standards bodies. Even within UMTS, there are variants of the standard. This book is intended to be practical in nature, and hence, I have chosen to focus only on the widely deployed frequency division duplex (FDD) mode of operation. All of the descriptions of functionality, the examples, the test requirements and so on are specifically only covered from the FDD perspective.

The purpose of this book is to provide an introduction to the complexities of testing a UMTS UE through its design cycle. Unless specifically stated, all the examples and descriptions represent the view from the terminal side. This includes testing done during the product development phase, conformance certification and gaining acceptance by operators for deployment on their networks.

The book is divided into three sections. Part I provides an overview of the following:

• Mandatory processes the UE has to go through
• Expectations of operators and end-users
• Typical testing done to ensure a high-quality product.

Where possible, I have also tried to cover some of the practical issues, such as how to go about testing, what equipment is typically needed and some of the common problems encountered during testing.

Part II provides a more detailed look at the testing of the main layers of the air interface protocols of the UE, starting with the physical layer and working up to the signalling and some of the system testing required. The chapters in this section are structured to provide:

• A basic introduction to the technology and protocol behind each layer
• An explanation of the test requirements associated with that layer.

Testing the higher layers is a very substantial subject in its own right, and looking at each of the signalling protocols in isolation does not help the reader to understand how the system works as a whole. Chapter 12 explains a number of complete signalling procedures, showing how the individual protocols work together in a structured way. These procedures are explained from the perspective of a test system and are intended to help the reader understand the conformance test cases and gain a starting point for writing test cases.

The book concludes with a brief section looking at some of the trends shaping the future of testing mobile UE.

While the text will hopefully provide useful background information on UMTS, the behaviour of the UE is specified over hundreds of thousands of pages of detailed specifications. In a book of this type, it is not possible to provide detail on every aspect of the UE specifications. Instead, the explanations are intended as an introduction only. They are incomplete in that they explain only at a high level, and many detailed points are omitted for clarity. The reader is recommended to refer to the full 3GPP specifications to understand the full detail of any individual function.

1.3 Overview of the Industry

The mobile communications industry has experienced one of the fastest growths in history. Since the introduction in 1981 of the Nordic Mobile Telephone system, the world's first fully automatic cellular system, the industry has grown to service over 2.6 billion subscribers in 2006; more than one third of the Earth's population. The industry inherits much of its DNA from the fixed telecommunications industry, and this has important consequences when considering testing. Considerable attention is paid to creating and adhering to open standards. Equipment suppliers are expected to prove conformance to these standards as well as their ability to interoperate with other suppliers. Equipment is expected to operate reliably and continuously for long periods of time.

These requirements combine to create some tough challenges for manufacturers. The need for open standards and close adherence creates a relatively slow evolution path for new technology compared to, for example, the computer industry. The effort and time needed to define these standards give rise to periods of slow evolution interrupted by large technology steps to 'catch up' with the latest advances. The evolution from GSM to UMTS is an example of such a step.

The problem for the industry is that the average user already has an expectation of performance and reliability set by the preceding technology. The new technology has to equal this expectation, even from the early days of introduction, otherwise it quickly gets a reputation for being unreliable, and this can act as a significant brake against uptake. This is a tough challenge indeed, as the comparison is generally being made against a technology that has had years of gradual evolution to become stable and robust. Against this backdrop, testing takes on a higher degree of significance.

Most organizations and businesses involved in the mobile communications industry are connected in some way with testing of UEs, either indirectly or directly as a key requirement, or even as their primary business focus. The next few paragraphs look at the impact of testing on the main parties associated with mobile communications.

1.3.1 Network Operators

Network operators provide a variety of wireless services based on a cellular wireless infrastructure coupled with a license to operate that infrastructure. The license usually comes from the national government and lays down conditions or regulatory requirements that the operator must meet, such as operating frequency bands and the radio technology of the network (although more recently the latter is changing). Originally, the network operators set up, operated and maintained the physical network themselves. However, more recently, some operators have been externally sourcing these, often from the network infrastructure suppliers.

In many areas, the network operator supplies a complete service to end-users, including the phone or other UE, creating a direct relationship between the operator and the end-user. This places the operator as the first point of call if the UE does not operate correctly. As well as loss of revenue from dropped or missed connections, the operator also has to deal with the consequences of problems, which may often result in end-users calling the operator's support call centre to complain. Consequently, operators have a strong interest in testing UEs, and many operate fairly extensive acceptance testing to try to catch problems before phones are supplied to subscribers. This is covered in Chapter 6.

1.3.2 UE Manufacturers

These are the companies that either undertake or commission the manufacturing of mobile phones, PDAs, data cards and various other mobile communications devices and market them to network operators and in some cases directly to end-users. The UE manufacturer is ultimately responsible for the quality of the device and therefore has the greatest interest in its testing. Even if the UE is made from components that have themselves been extensively tested, some level of testing is still required when integrating to a complete product, and some types of testing, such as conformance and acceptance testing, have to be carried out on the final device.

1.3.3 Component Suppliers

For all but the largest manufacturers, developing a mobile phone entirely through one's own resources is not practical. Many manufacturers rely nowadays on third parties to supply key components of the phone. This has created a competitive market for chipsets and protocol stacks, and competition has resulted in pressure for greater integration. Nowadays, apart from vendors supplying specialized components, suppliers of UE chipsets are expected to supply a complete solution, including a reference design capable of passing certification and accompanying protocol stack software. Whilst the final integrator of the end product still needs to perform a certain amount of testing, there is an expectation that this testing will be minimal or perfunctory, with main design verification being done by the component supplier. This means that component suppliers will also have an interest in carrying out development, conformance and interoperability testing.

1.3.4 Testing Services

A number of specialist companies provide testing services to the industry. Mainly, these are targeted towards the UE manufacturers and component suppliers although increasingly they are also supplying the network operators. In particular, certification of UEs requires specialized test laboratories that meet certain quality standards, usually assured through a system of accreditation. These laboratories, also often called 'test houses' in many cases, are starting to expand their business by offering a wider range of test services. Traditionally, these companies have focused on offering conformance testing. However, there has been a steady trend for them to become more involved in operator acceptance testing and other forms of interoperability testing.

1.3.5 Standards Bodies and Certification Bodies

Standards bodies define the behaviour and operation of the network technology at least to the point where independent implementers should be able to develop devices that work with each other. This usually includes providing a test specification that sets out what tests need to be performed to ensure that a device meets their specification. A certification body is one empowered with the authority to decide whether a device meets a minimum level of compliance with the standard. Their authority can come from national government or it can come from the industry itself, through self-regulation. Standards bodies can sometimes be separate from certification bodies, such as is the case for UMTS – where the standards body is 3GPP and for Europe, the certification body is the Global Certification Forum (GCF). There are many reasons for this independence; for example, deciding the minimum acceptable level of compliance is often a local or regional affair, whereas defining the standard may be a global one.

In the past, standards bodies have only been concerned with conformance testing, but there has been a trend more recently for an informal involvement in interoperability testing through the hosting of interoperability test events. Certification bodies by contrast have some history in both conformance and interoperability, particularly in areas where the industry considers that conformance testing on its own is not enough to ensure adequate quality.

2

Introduction to UMTS

2.1 The History of UMTS

The third-generation mobile systems have their roots in a project set up initially in 1985 by the ITU called International Mobile Telecommunications 2000 or IMT-2000. The '2000' represented the concept that this was communications in the new millennium, as well as an anticipation that the frequency band for operation would be in the 2000-MHz range with data rates of 2000 kbps. This project gained momentum in the wake of the 1992 World Administrative Radio Conference, held by the ITU to look at globally harmonized radio spectrum allocations across a number of areas, including mobile telecommunications. The principal goal of the project was to try to identify a band of radio spectrum that could be used in as many regions of the world as possible to promote the concept of a truly global system. This is a little more complicated than it might first appear. In order for governments to allocate or free-up spectrum, they need to know the implications for other spectrum users, particularly in neighbouring bands. This requires a degree of RF specification in terms of creating a framework within which the new systems could sit. This work was actually completed in May 2000 when the interface specifications were finally ratified by an ITU assembly meeting, but by this time, the project had already gathered considerable pace and the work to define mobile telecommunication systems that fitted into this profile was already well under way.

The IMT-2000 project identified five potential radio interfaces, based on various combinations of three fundamental technologies: CDMA, time division multiple access (TDMA) and frequency division multiple access (FDMA). The ITU then invited bids to provide standards for each of the interfaces, with a goal of providing a closely related

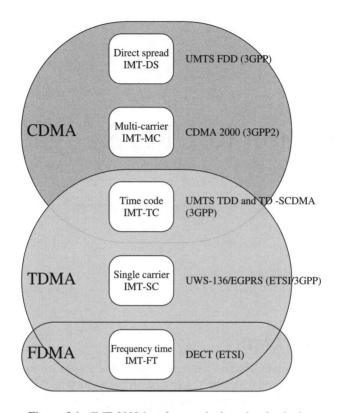

Figure 2.1   IMT-2000 interfaces and adopted technologies

family of specifications. The interfaces and their adopted standard (Figure 2.1) were as follows:

1. Direct spread (IMT-DS): Also known as direct sequence, this is where different channels or data for different users are multiplexed using orthogonal codes and then spread over a wide frequency using a pseudorandom sequence. This is the technology that ETSI selected for the evolution of the GSM network, which became known as UMTS, and that Japan selected to replace its overloaded second-generation networks.
2. Multicarrier (IMT-TC): This is basically a variant of direct-spread CDMA but is tuned towards the evolution of the existing CDMA system operated in the United States, Japan, Korea and other parts of the world. The older CDMA standard (known in the United States as IS-95) operates in a 1.25-MHz bandwidth. The evolution of this aggregated up to three of these carriers to provide data rates that reached the IMT-2000 goals. This evolution is known as CDMA-2000 and also as 3XRTT (for the three carriers).
3. Time code: This is a combination of direct-sequence CDMA but also includes the concept of dividing the air interface into timeslots. The main purpose of this is to use up unpaired spectrum. Most cellular communications systems use separate frequencies for

each communication direction. Having the transmit and receive frequencies reasonably well separated helps to keep the radio design complexity down. However, this requires pairs of frequency bands and hence a wider chunk of spectrum. The WARC conference identified some unpaired bands which were available but required a different approach. The main goal of the time-code approach was to specify a system that can allocate timeslots for use either as transmit or as receive paths and therefore make optimum use of these unpaired bands. This gave rise to a proposal based on introducing a timeslot arrangement into the physical layer for the UMTS proposal, which became known as time division duplex (TDD) mode. The Chinese standards organization China Wireless Telecommunications Standard (CWTS) also had a proposal for a time division system based on CDMA, but at a lower spreading rate and bandwidth, which became known as TD-SCDMA. In the end, the two proposals were merged together to create high chip rate (HCR) and low chip rate (LCR) modes of operation to complement the TDD proposal.

4. Single carrier: The United States also had a number of networks using a TDMA system called IS-136. An extension of this system based on ETSI's EDGE standard was selected and proposed as part of the IMT-2000 family. This was officially called IS-136 HS but also became known as EDGE Compact.

5. Frequency time: The IMT-2000 programme also developed the concept of 'domains' that networks could focus towards, such as terrestrial, satellite and so on. The terrestrial domain was divided into 'indoors' and 'outdoors'; the meaning here being whether the network was intended to cover a broad geographic area or be restricted to within a building or campus. In the main, the other proposals were focussed on outdoor operation although UWC-136 also had an indoor mode. An existing European standard from ETSI called Digital Enhanced Cordless Telephony (DECT), which had been developed as an indoor system, was proposed to meet the IMT-2000 indoor component. DECT used a combination of time division and frequency division, with dynamic allocation of frequencies, an approach that is difficult for an outdoor system where frequency allocations have to be carefully planned. The proposal was to extend DECT from its original capability of around 700 kbps to a target of 2.5 Mbps.

From the perspective of UMTS, it is the IMT-DS selection that we will focus on, but it is worth noting that the IMT-TC proposal also has some bearing. For a number of years, the European Commission (EC) had been sponsoring research into next-generation mobile communications through its research of advanced communications technologies in Europe (RACE) and Advanced Communications Technologies and Services (ACTS) programmes, and this had resulted in a number of direct-spread CDMA projects. In parallel, the Japanese industry, mainly sponsored through an experimental network developed by the major Japanese operator NTT DoCoMo, had also done a significant amount of DS-CDMA research, resulting in a working prototype network. The key Japanese standards organizations, the Association of Radio Industries and Businesses (ARIB) and the Telecommunication Technology Committee (TTC), started a close cooperation with ETSI, which resulted in the European and Japanese proposals merging together to create a single, combined approach. The IMT-TC bid, based on a TDD version of this approach, also brought about a cooperation with CWTS and the Chinese industry. Based on the relationships that ETSI was fostering, a partnership of standards organizations started to come together. This became known as the 3rd-generation partnership project, or 3GPP.

2.2 The 3GPP

Formed in December 1998, the 3GPP brought together five regional standards organizations:

- ETSI, and in particular the Special Mobile Group (SMG) committee
- The Alliance for Telecommunications Industry Solutions (ATIS) from the United States, and in particular its T1P1 committee
- The ARIB from Japan
- The TTC, also from Japan
- The Telecommunication Technology Association (TTA) from Korea.

In May 1999, the Chinese standards organization China Wireless Telecommunication Standards Group (CWTS) also joined as the sixth organizational partner. CWTS has now become a part of the China Communication Standards Association (CCSA), which replaced CWTS as an organizational partner in May 2003.

The work already starting under ETSI's SMG was transferred over to 3GPP. The operating procedures and general committee structures were largely inherited from ETSI's way of working. For a while, ETSI's SMG and 3GPP continued to develop GSM and WCDMA in parallel, but the close relationship between the two standards activity made some sort of a merge inevitable. In July 2000, the future maintenance of GSM was wholly transferred to 3GPP and the SMG ceased operating. Instead, a new group within 3GPP, known as the GERAN, was formed.

2.3 Organization of 3GPP

The full structure of the 3GPP Technical Specification Groups (TSGs) is shown in Figure 2.2. There are four specification groups, each consisting of a number of working groups (WGs). Between them, they are responsible for over 2200 3G specifications and 1300 GSM specifications. The suite of specifications contain core specifications that define the operation of the various parts of the network or the protocols used in the network and test specifications that detail how implementations will be tested to ensure they conform to those core specifications. The main groups involved in writing and maintaining conformance specifications are TSG GERAN WG3 for GSM and TSG RAN WG5 for UMTS. Interworking between the two radio access technologies (RATs) is handled jointly, with RAN WG5 responsible for mobility from the UTRAN to the GERAN and GERAN WG3 responsible for mobility in the other direction. SA WG3 and CT WG6 also cover some specific areas of conformance testing. These groups and their respective specifications are discussed in more detail in Section 5.2.

Many of the other groups produce core specifications that either directly or indirectly have an impact on the testing of the UE, but the major groups of interest, together with their key specifications from a UE test perspective, are summarized below:

RAN WG1	Responsible for the Layer 1 series 25.211–25.215 (FDD)
RAN WG2	Responsible for the Layer 2 specifications 25.321 [media access control (MAC)], 25.322 [radio link control (RLC)], 25.323 [packet data convergence protocol (PDCP)] and the radio resource control (RRC) specification 25.331
RAN WG4	Responsible for the UE radio specification 25.101
CT WG1	Responsible for the layer 3 Non Access Stratum (NAS) specification 24.008

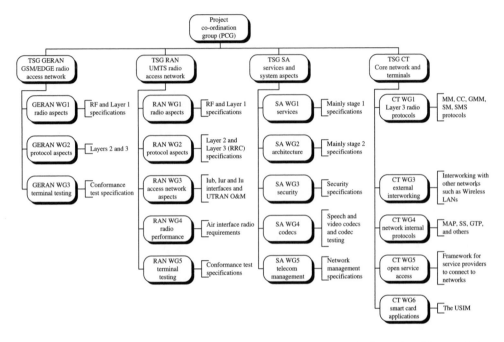

Figure 2.2 Organization of 3GPP technical specification groups and WGs (reproduced by permission of ETSI)

2.4 Goals and Achievements

In 1998, DoCoMo, the largest operator in Japan, established a WCDMA trial network based on many of the underlying concepts in the joint European and Japanese submission for IMT-2000. Capacity pressure on the Japanese second-generation networks created strong pressure for an early deployment of the third-generation technology and, on 1 October 2001, DoCoMo launched the first commercial network based on the UMTS WCDMA standard.

By June 2001, the 3GPP specifications were becoming reasonably stable, and early in 2002, the first dual-mode (GSM and WCDMA) terminals were becoming available to support the network trials that were starting across Europe. Among the trials, a few operators decided to launch their networks even though terminals were not yet ready for general resale. They included Telenor of Norway and Mobilkom of Austria. The first major UMTS network launch in Europe, however, came from the operator '3' in the United Kingdom, who were unusual in that they did not have an existing GSM network. In keeping with their high-profile approach, they officially started commercial operation on 3 March 2003, using mainly terminals and infrastructure very similar to those already in operation on DoCoMo's 3G network. UMTS networks are now in operation in both East and West Europe, Japan and many parts of southeast Asia, the Middle East, South Africa, and North and South America. Already some networks have put high-speed packet access (HSPA) into operation, offering speeds of 14.4 Mb s^{-1}, over seven times the target rate envisioned by IMT-2000.

3

Types of Testing

Modern mobile phones are very complex systems containing a number of quite substantial subsystems. For example, a standard mobile phone may contain an audio subsystem, a telephony subsystem, a camera, a packet switched modem, a web browser and Internet protocol stack, a java engine and so on. Each subsystem needs to be tested to make sure that it works in its own right, but the system needs to be tested as a whole. That is, the interactions between subsystems – intended or incidental – also need testing. This gives rise to a number of different testing strategies and methods that are used throughout the process of developing UEs.

3.1 The Purposes of Testing

Ultimately, the aim of all testing is to make sure that the device works correctly. Of course, the ideal is to fully test everything within the UE, but in practice, this is not really possible. Just considering the interactions between subsystems, if a UE has 10 subsystems then to test all combinations of subsystems operating together would require 1023 sets of tests. If each subsystem can do 10 different things, the total number of tests would be almost 10^{10} and that is just to cover interactions. Accepting that exhaustive testing is not really practical, we can redefine the aim of testing as finding the optimum balance between two opposing constraints:

1. What is a reasonable cost in terms of resources? and
2. What is an acceptable risk of letting through a bug or defect?

This definition now opens up a world of possibility. Testing is a very resource-intensive business, in terms of both manpower and specialized equipment. The cost that this carries eventually filters its way down into the cost of the UE, albeit often indirectly. The number

Testing UMTS: Assuring Conformance and Quality of UMTS User Equipment   Dan Fox
© 2008 John Wiley & Sons, Ltd

of bugs that get through into the final products bears a direct relationship to the end-user's perception of quality. Figure 3.1 shows at a conceptual level how we might view the impact of increasing the amount of testing on product quality. The wide base of the pyramid represents the fact that a set of basic tests are going to catch many fundamental problems. As the amount of testing increases, the number of problems caught will decline, simply because the earlier testing will catch a number of systematic issues. By the time we reach the level of testing usually required by certification programmes, there should be no basic issues left. However, there may still be more complicated bugs, especially ones arising from interactions between multiple subsystems. In theory, testing could continue towards the top of the pyramid to provide 100 % coverage, but in practice, the amount of coverage has to be traded against the time taken to both develop and perform the testing. From this definition, testing now has a relationship to the overall business strategy. For example, a company that wants to position itself as a high-quality supplier will need to invest more heavily in testing and accept that the quality image will create enough customer demand to allow them to maintain a price premium. A company focussed on the low-cost end of the market will usually have to set the balance point so as to keep testing costs lower. This picture is complicated by the fact that the larger manufacturers can spread the cost of their test programmes across their total product range using platform development techniques, so that their low end products might be able to enjoy higher levels of testing than could be justified if they were developed in isolation.

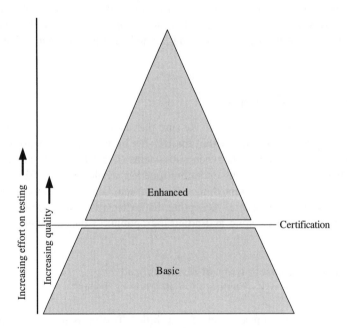

Figure 3.1 Product quality and testing

What may appear as a simple question turns out to be a complex issue of balance that relates ultimately to business strategy. However, manufacturers do not have a completely free hand in deciding the balance point. In a sense, the mobile communications industry grew from fixed wire telecommunications, and particularly in the early days, it was subject to fairly extensive regulatory requirements in most regions of the world. Despite the more recent deregulation that has accompanied the globalization of cellular communications, the industry still has firmly embedded in it the concept of a minimum acceptable set of requirements that devices have to meet.

3.2 The Expectations on an Equipment Developer

We can broadly separate testing into two types: testing that is done as part of the development of the UE, including testing that is done at the end of the development to prove that the design is conformant to the relevant standards, and testing that is done postdevelopment, once the UE has been commercially released. This book is predominantly about the former of these, as this area makes up the major burden of testing; however, we will look briefly at the testing done during the production life of the UE. There are various types of testing that have to be considered during product development, and the following sections take a closer look at each one. But in this section, each is introduced and its importance to the developer is discussed.

I have grouped these into three general categories:

1. Integration testing
2. Conformance testing
3. Interoperability testing.

Integration testing is the internal concern of the manufacturer, and as mentioned in the previous section, this can be core to the product strategy. This is the area where manufacturers have most discretion, but performing less integration testing can increase the amount of work and time needed to perform conformance and interoperability testing. In general, problems found in either of these areas can be far more difficult and costly to fix than if they are found during the development activity.

Conformance testing is generally a fixed quantity; the amount of testing is determined by the standards body. It is an essential part of the various certification programmes, and at a simplistic level, if the UE cannot pass the conformance tests, then it is not really conforming to the standard. In practice, many manufacturers use the conformance tests within their development for two main reasons. First, by continually running the conformance tests during development, the additional effort needed to get the product conformant at the end of the development is minimized, and second, the conformance tests are excellent and readily available source of high-quality tests that can be used to exercise a feature within the UE. Test systems will usually provide a great deal of logged information from running a conformance test, and it is often as easy to debug from this as from a custom-written test.

Interoperability testing is another variable quantity. It is difficult to do because it usually involves testing in a different location to the development, and this presents its own logistical problems. However, it gives the highest level of confidence that the UE will interoperate with a specific configuration of infrastructure. Naturally, network operators are very keen to know that the terminals they offer for resale will work against the network infrastructure they use, but there are a number of infrastructure suppliers, and with standardized network equipment interfaces, the number of possible permutations is large. The amount of interoperability testing is again partly a strategic decision. The more the configurations tested, the more operators the end product can appeal to.

To this list we can add one further type of testing; this is sometimes referred to as regression testing. The word regression means going backwards, and the purpose of the testing in 'regression testing' is to prevent the design from going backwards. This is the type of testing aimed at making sure that errors do not start creeping into those functionalities that have already been debugged. Regression testing is a very important part of keeping a development moving forwards, and building and maintaining a good suite of regression tests can contribute significantly towards higher quality and shorter development times. The tests themselves are usually drawn from the three areas mentioned previously; a selection of development, conformance and even laboratory versions of interoperability tests. The keys to successful regression testing are keeping the suite maintainable and the level of testing manageable. These are discussed further in Section 4.3.

3.3 Differences with Other Markets

The mobile communications industry inherits many of its genes from the telecommunications industry, where products are extensively tested and a great deal of effort is put into ensuring devices a conformant to the relevant standards and interoperate correctly. This is driven by a number of factors, among which are the clear lines of responsibility back to the network operator, such that a poor quality of service will inevitably lead to a loss of revenue both in terms of subscribers moving to other suppliers and in terms of losing the revenue from the service that failed. A second major heritage of the industry is from the radio side. Mobile operators provide their services using radio spectrum licensed by regulators, and as a result there are (or at least have been in the past) legal requirements to ensure a certain level of performance and quality of equipment particularly with respect to the radio parts. These requirements do not exist or are much weaker in other high-technology industries. For example, the use of formal testing is quite rare in the computer industry even though the Internet involves a wide variety of protocols. Here, testing tends to be predominantly discretionary and often consists of a combination of quality-focussed product testing and interoperability testing by trying the device in a number of realistic operating environments, usually against far-end equipment from other manufacturers.

3.4 Testing Through the Life Cycle

Nowadays, it is widely recognized that the earlier in the development cycle that testing can begin, the better is the result. The more complex the system, the more important this is. Bugs, or faults, left in the code tend to get buried as more code is added to the system. When they finally are uncovered, they can precipitate a waterfall of dependent changes that are required to the subsequently added code. This is very time consuming and can be a significant cause of unpredictability in development project timescales.

By testing thoroughly from early in the development, these occurrences can be minimized. This is a key part of most modern development methodologies, such as the Waterfall and V-models (Figure 3.2).

A modern mobile device contains a number of semi-independent subsystems, in many cases requiring different engineering skills to build. The overall development will often

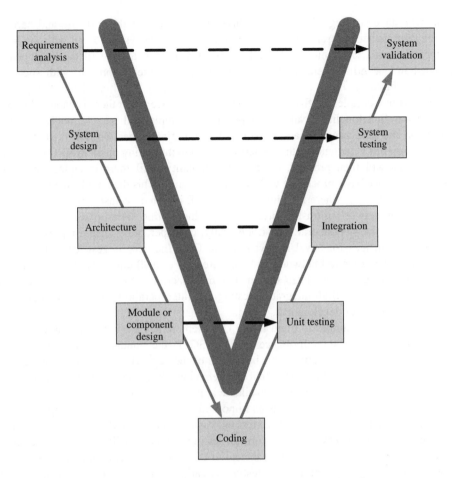

Figure 3.2   Classic V-model for the system development and testing process

involve a number of smaller teams with different disciplines, working in parallel strands that each go through their own development cycle before coming together in a superintegration exercise. In this chapter, I have summarized the different types of testing applied through the development process. The subsequent chapters provide more details on each type.

Testing usually begins at the unit or module level. In software, this involves the development of test frameworks that test at the interfaces of modules. Generally, testing at this level is a software only affair and requires no specialized tools. Test cases are highly specific and also dependent on the module design and system architecture. In hardware, this usually involves general purpose test equipment, such as signal generators or arbitrary waveform generators and signal or spectrum analysers. While it is possible on many high-end signal generators and analysers for the user to generate (or analyse) signals from mathematical models, using Matlab for example, this is often not necessary. Equipment manufacturers usually supply software or built-in functions, even from quite early stage in the life of a communications standard, and this would be suitable for most development purposes.

The next stage is integration testing, where at least parts of the software and hardware are brought together and run for the first time as a complete system. Even with the largest development team, it is unlikely that all the functionality of a UE will be developed and integrated in one go. Generally, development takes a phased approach, with successive integrations building upon a stable base of functionality resulting from the previous integration. For example, the first integration may take place when there is enough hardware and software functionality available to perform a simple cell search, read the system information and send a random access request for an initial connection. The next integration might build on this by extending the scenario through to the setting up of an RRC connection.

Integrations will then proceed, building functionality until there is enough present to support at least one basic procedure with real infrastructure. This then marks the beginning of interoperability testing, where functionality in the UE is tested against various combinations of commercial network infrastructure. Usually, this testing takes place on special test networks inside shielded laboratories. Logistically, interoperability testing represents one of the biggest challenges for any UE development. Only the very largest manufacturers can afford to have their own test network, and there are only a small number of these laboratories in the world that are available to the smaller manufacturer. As a result, the demand for laboratory time is high. Both integration testing and interoperability testing continue for the duration of the product development.

As functionality continues to build in the UE, there comes a point when there is enough to be able to run a simple conformance test. The 3GPP, the standards body responsible for defining UMTS, as well as defining the behaviour of the network and UE, has also defined a large suite of tests intended to ensure that UE behaviour does conform to the standard. This suite of conformance tests is publicly available and supported on a range of test equipment from several manufacturers and forms an important reference for how a UE implementation should behave. In the past, conformance testing was often considered as a function at the end of the development process. Like any form of testing though, it will throw up development errors, and the earlier these are caught, the easier they are to fix. Therefore, nowadays there is a strong drive to start conformance testing as early in the development as possible and then to test continuously through the rest of the development process.

Integration, interoperability and conformance testing continue throughout the UE development, with new tests added as functionality grows. In the best development practices, a broad selection of key tests of all three types are collected in a test suite and used for regression testing. That is, each new software build is tested against this suite to make sure that the existing functionality has not been broken by any new functionality or recent bug fixes. Regression testing forms a key pillar of any good development and is covered in more detail in Section 4.3.

Towards the end of the development, two important test hurdles arise. These are gates to the commercial deployment of a product and in theory should be an extension of the testing already done. The first gate is device certification – that is, independent verification that the UE passes the necessary conformance tests. The second gate is operator acceptance. In most regions, the supply of the phone to the end-user is a part of the service provided by the network operator, and therefore the only practical way to sell a product in reasonable volumes is for it to be promoted by the operator. Many operators have a process for accepting devices, and this can include a regime of detailed testing both in the laboratory and on the live network.

Once a UE has passed its certification, has been accepted by some operators and has passed whatever internal milestones are required (e.g. pilot production runs, internal verification testing and quality assurance), it is generally ready for commercial release. Depending on the target markets, this can generally be a rather soft end to the development. Once a device gets into the hands of the general public, it is usually not practical to change or update the software in any way. However, for a successful model, usually as more operators take up the UE, they can sometimes impose their own specific requirements, which require software modification.

Beyond commercialization, the focus of testing mainly switches to manufacturing. Here, the main purpose is to test those areas which are subject to variation or unreliability due to the manufacturing process. This relates mainly to the RF transmitter and receiver, with perhaps a confidence test of some key subsystems acting as a final quality gate.

4

Integration Testing

4.1 Definition

Integration testing in this context refers to a range of internal testing that is done throughout the development process. It follows unit and module testing, with some area of overlap as the various parts of the system are built up. It mainly refers to the testing done as the software and hardware are first brought together and runs through the development as functionality is added to the system piece by piece such that significant parts of the system functionality are achieved. It covers full system testing, where all the parts are tested as a single unit, checking complete vertical paths through the applications, protocol stack and physical layer. To keep integration testing distinct from the other types of testing, it can be characterized as very much tuned to, or targeted on, the function being debugged or integrated. Integration tests are not aimed at finding general problems in the design, and in fact tests that pick up other problems can be undesirable. These problems may already be known about and under investigation in another part of the team. Tests also may need to be capable of rapid adaptation as engineers often want to vary a test repeatedly as they try to narrow down the root cause of a problem.

4.2 Getting Things Working

Working up the right-hand side of the V-model in Figure 3.2, the first stage of module testing consists of two parts:

1. A classical software test part, which can be addressed by developing unit test frameworks that allow the external interfaces of each module to be exercised. This type of software testing is beyond the scope of this book.
2. A hardware test part, which is quite a specialized task and can often use the same types of equipment as the later integration stages.

Testing UMTS: Assuring Conformance and Quality of UMTS User Equipment Dan Fox
© 2008 John Wiley & Sons, Ltd

Radio testing is discussed in more detail in Chapter 8, but at a high level radio module testing is usually performed using discrete test instruments: a vector signal analyser for looking at the output of the transmitter and a vector signal generator (VSG) for driving complex and realistic signals into the input of the receiver.

The baseband processing is generally developed in an application-specific integrated circuit (ASIC), and conventionally this is tested through software simulation during development and then using logic analysers prior to integration with the radio. However, there has also been a trend towards the use of ASIC simulators. These are flexible prototyping systems that map the functionality of the ASIC onto a set of field programmable gate arrays (FPGAs). The resulting emulation behaves like the final ASIC, can be quickly reprogrammed to fix any errors found during testing, but due to the added delays of getting signals between devices, tends to run much slower. These ASIC emulators can be tested using a system simulator, but it needs to have the capability to connect signals from its digital baseband interfaces and to run at reduced speed. System simulators with this capability have been available since the early stages of UMTS but are still the exception rather than the rule.

In parallel with the integration of the physical layer, work can be done independently integrating the various modules from layer 2 upwards. A number of virtual or host test solutions are available to support this. These solutions replace the actual layer 1 with a virtualized implementation as shown in Figure 4.1. Typically, a virtual tester will run

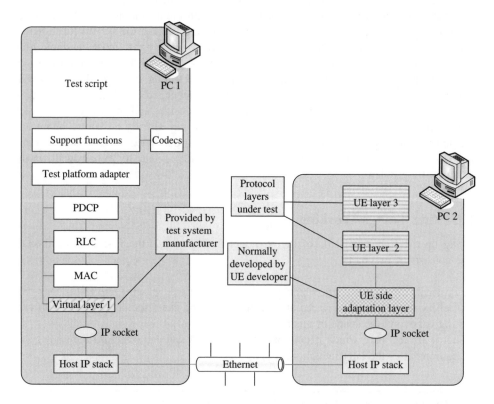

Figure 4.1   Virtual test environment for testing UE protocol layers without target hardware

on a Windows PC and connect using TCP or user datagram protocol (UDP) sockets. Frames of data are relayed over the sockets. In the simplest case, these may just contain an identification of the logical channel identity and the transport block content. In more sophisticated simulations, this will contain a more complete description of the layer 1 frame. At the far side of the link (PC 2 in the figure), the UE stack developer writes an adaptation layer which converts between the framing information provided on the socket, which is often described in XML, to the proprietary interface primitives used to communicate between his layer 2 and layer 1. The virtual test solutions provided by the system simulator manufacturers are usually compatible with their hardware test solutions so that the test scripts developed and used in this environment can be reused during software–hardware integration.

Once the integration of the protocol stack and the physical layer starts, there is a need for a specialized piece of test equipment, known variously as a system simulator, protocol tester, network simulator, basestation simulator or signalling tester. Generally, in this text, the term system simulator, or SS, is used. The SS is described in more detail in Section 11.1. In simple terms, the SS consists of two parts: a radio modem, supporting the physical layer and layer 2 of the network side of the air interface, and a test script engine, capable of running scripts which both control the radio modem and transmit signalling and data to the UE. The radio modem is similar to the layer 1 of the Node B but differs in two important aspects: it is only designed to communicate with a single UE and the radio section is designed to transmit the radio signal only over a very short distance, or more typically using a coaxial cable connected directly to the UE antenna port. As a result, SSs are not licensed radio equipment and can be operated in test laboratories without any special considerations.

SSs are programmed through test cases or test scripts. These are sequences of commands that either:

- Configure the radio modem
- Send signalling messages or data to the UE
- Receive signalling messages or data from the UE
- Provide programmatic support, such as manipulation of variables, flow control and so on
- Call other sequences of commands or other programs.

These test cases are used to provide simulation of higher layer (layer 3 and above) network signalling and allow the protocols in the UE to be thoroughly exercised.

Integration requires a slightly different type of test case than other types of testing. The main difference arises from the fact that much of the later verification testing of the UE, such as conformance testing, assumes that the majority of functionality in the UE basically works correctly, and the test can focus on checking one specific aspect of behaviour. However, for integration testing, this assumption is not valid. Particularly in the early stages, it is often better to assume that the functionality does not work in totality, and the aim of testing is to establish which areas do work. This results in some significant differences in test case design, with tests generally being simpler and written to assume as little additional functionality outside of the area of test as possible. By contrast, even basic conformance tests are written assuming fairly significant functionality is already present in the UE.

A well-designed library of integration tests will go through a series of integration milestones, each carefully building on the functionality already tested in the previous milestone. A simplified view of these milestones might be as follows:

- Broadcast of system information
- Initial location update for registration using CELL_FACH state (see RRC States, Section 10.2.2)
- Location update using CELL_DCH state
- Mobile-originated/mobile-terminated speech calls in CELL_DCH state
- Cell reselection and Public Land Mobile Network (PLMN) reselection
- Packet data protocol (PDP) context activation
- Transitions between CELL_DCH and CELL_FACH
- Hard handover
- Soft handover
- Inter-RAT reselections
- Inter-RAT handover.

As functionality builds in the UE, other types of test can be added into the test regime. Once the UE is capable of a certain level of basic functionality, testing can start against infrastructure equipment, and once it is capable of running the first conformance test cases, these can become an important part of the further integration and verification process.

4.3 Keeping Things Working

An important part of any development process is to keep the development moving forwards. It is very easy when dealing with complex projects for a fix to one area of the code to break something in another area. If these breakages are not found quickly, they can become buried by further functionality and cause much greater problems down the line. This can be done by regularly testing the evolving software using a wide variety of tests. This process is known as regression testing.

It is generally recognized that the more frequently regression testing can be performed and the more tests that can be run, the better. Often regression testing performs a gate so that further development work does not begin on a new build of code until it has passed the regression test suite. In general, the larger the development team, the more frequently the work of the individual engineers and teams needs to be merged into a single build and hence the more frequently regression testing needs to be done. In some cases, this can be as often as two or three times per week. This brings us to the major challenge of regression testing: time. A regression suite needs two main components. The first is a set of functional tests that perform a reasonably detailed check of all the main functions so far integrated into the UE software. These can be mainly success scenarios, but a selection of negative, or adversarial, scenarios, including rejected procedures, rejected calls and radio link failures, can also be useful. Second, the suite needs some stress tests. These are tests that perform large numbers of repeated scenarios, such as hundreds of calls in succession, often including some test cases at the boundary of the UE's capabilities, such as the maximum data rates, or radio bearer configurations using large buffer sizes and long timers, so that internal queue management is exercised.

However, as the complexity in the UE builds, the number of tests in the regression suite must also grow, and the time taken to perform regression testing becomes a significant bottleneck. For example, a voice call can take several minutes from start to finish, so a loop of 100 voice calls can take four or five hours to run. A reasonable regression suite can take several days to run through, and very soon this task becomes impractical unless the process can be automated.

4.3.1 Automation

Many SSs include the ability to embed AT modem commands into the test script and to have these transmitted to the UE as part of the test sequence. The test script engine in the SS is often based on a standard PC, and the normal modem driver for the UE will allow easy interconnection between the simulator and the UE's modem port using a USB cable, infra-red or even Bluetooth if the UE and the PC are so equipped. AT commands allow a certain amount of automated control during test operation and can perform many common triggering functions in the UE, such as initiating a voice call, answering a mobile-terminated voice call, initiating a PDP context activation, deactivating a PDP context and even in some cases more sophisticated operation such as performing manual network selection. However, the main limitation of AT commands is that they cannot power the UE on and off. This is one of the most frequently required operations in test cases, and a variety of solutions are used:

- Proprietary-dedicated electrical interfaces: These are often used during development to connect external logging equipment to the UE and for various other maintenance functions. Such interfaces are custom designed to provide fast and comprehensive automation support and are disabled or removed on commercial versions. Being proprietary, these interfaces require custom software to interface to the SS.
- Mechanical methods: These are methods where commands from the simulator are used to drive actuators, or even robot arms, to physically press buttons on the UE keypad. There are various mechanical jigs available commercially, which can be adapted to different phone shapes and sizes, but nowadays, the huge variety of phones means that there are always new formats that the jigs cannot handle. Mechanical jigs are usually fairly simple to interface to. Simulators often provide semicustomisable interfaces, where commands sent from the simulator are translated using a user-supplied look-up and sent to a serial or parallel port to which the jig is connected.

Another method of automation useful when a number of UEs with different form factors are being tested in parallel is simply to remove the keypad and solder leads or force contacts directly onto the printed circuit board (PCB). Key presses can then be simulated with electrical switches or relays controlled directly from the SS.

Regardless of the method, automation is a key factor in effective regression testing, allowing tests to be run for long periods without operator intervention and allowing scalability so that the test load can be spread over a number of test systems, each running a different part of the overall regression suite.

It is worth noting that automation has become much more important for UMTS. The standard is much more complex than its 2G predecessors, and the amount of testing has consequently increased significantly. This has also placed new constraints on SSs, making the stability of the instrument a much more important factor than it was in the past.

5

Conformance Testing

5.1 History

Conformance testing within the telecommunications industry started out as a key part of the official type approval regimes that many national telecommunications authorities had in place when the industry was heavily regulated. During the deregulation of the industry in the late 1990s, type approval has been largely replaced by self-certification. The types of tests and the general level of testing have remained the same, but the regimes imposed by national authorities have been swept aside and replaced by another system of the industry's own choosing. One important consequence of this change is that it has catalysed a far more common approach to type approval throughout the regions of the world.

Modern mobile cellular communications devices inherit their type approval requirements from two historic sources: radio conformance and telephone conformance. Most regions have radio conformance requirements dating back to the start of licensing of radio spectrum by the authorities (1904 in the United Kingdom, 1925 in Japan). Radio conformance has been mainly associated with correct use of the radio spectrum, and control of interference with other users, but more recently an element of public safety has also come in. Telephone type approval regimes, which also date back a long time, are mainly concerned with protection of the network equipment and with establishing a minimum acceptable level of service from end-user equipment.

The following table summarizes a typical set of type approval requirements on a cellular telephone:

Test area	Radio	Telephone	Unique
Radio receiver and transmitter RF performance characteristics	☑		
Signalling protocol conformance		☑	
Electromagnetic radiation compliance (EMC)	☑	☑	
Audio performance		☑	
Absorption of electromagnetic radiation by the human body – specific absorption ratio (SAR)			☑

Testing UMTS: Assuring Conformance and Quality of UMTS User Equipment   Dan Fox
© 2008 John Wiley & Sons, Ltd

5.1.1 The Importance of Different Type Approval Areas

The radio-related type approval areas tend to have a regulatory backing, even in these days of deregulation. They are generally of importance to the radio spectrum regulator and are the basis for ensuring the license that has been issued is properly followed or used. As a result, they tend to be regarded as the most critical tests in the type approval regime. However, there are not so many radio parameters to test, so in terms of the number of test cases that need to be checked, they form only a small part of the total. The proportion of each area for the UMTS certification tests is shown in Figure 5.1.

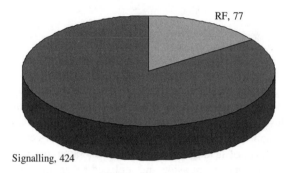

Figure 5.1   Relative proportions of RF and signalling test cases for 3GPP Release 99 (based on GCF Work Items 10 and 12)

Signalling conformance makes up the bulk of the tests by number, and is important to the network operator. The advent of Integrated Services Digital Network (ISDN) in the 1980s introduced digital signalling between the user and the network, and this resulted in a large step up in complexity. When cellular networks became digital with the second-generation standards, this complexity grew further as terminal signalling now had to handle mobility in addition to the services already established for wired digital terminals. It is not surprising then that many of the principles established for ISDN type approval have persisted, and been built on as the basis for cellular telephony type approval.

5.1.2 The Origins of UMTS Conformance

At the time when the UMTS standardization activity was kicking off, GSM was already established as the most widespread and successful commercial cellular standard. Of all the second-generation standards, GSM had the best developed formal type approval system, based on a rigorous set of conformance test cases developed by the standards authority, ETSI. The relatively smooth interoperability, across a large diversity of network deployments, made GSM the obvious model to follow for UMTS. As a result, ETSI was quickly thrust into a leadership position in developing the conformance tests for UMTS.

Many of the companies involved in, and represented at 3GPP, also had a long history of GSM and involvement in its standardization. They brought with them a good deal of knowledge of what worked well for GSM, and which approaches were not so successful.

There was a concerted effort within 3GPP to reuse the best practices of the past, but also to learn from the mistakes. An example of this is the creation of a TSG with the specific aim of developing the conformance tests for the UE, following the example of ETSI's SMG-7. In the case of 3GPP this group was established almost in parallel with the core specification groups. Normally, it is considered difficult to develop conformance specifications before the core specifications they are testing conformance to are stable. Test writers can waste a lot of time and effort testing requirements that subsequently are removed. However, there are also some significant benefits: the availability of conformance tests can be accelerated; and the test writers can give valuable feedback to the core specification developers on features or requirements that are difficult to test. Whilst earlier availability of tests is very difficult to judge as there is no easy benchmark, it is widely accepted that to some degree both benefits were realized, and the approach taken by 3GPP is now itself looked on as the model for future standards, and is being closely followed for UMTS long-term evolution (LTE).

Another example of learning lessons from GSM was the approach taken to the development of executable test cases. In other second-generation standards, the standardization forum usually stops at the development of a written (or prose) description of the test cases. In GSM a mixture of approaches were taken. For the RF tests, due to the nature of the tests, implementation was left to the test equipment manufacturers. For signalling tests, originally, ETSI set out with the intention of implementing these tests within the standards body, producing both a written description and a reference implementation that could be executed fairly directly on test equipment. Although this effort eventually transferred largely to the industry, under funding from the GSM Association, it resulted in a large body of publicly available conformance tests written in the standardized test language, Tree and Tabular Combined Notation version 2 (TTCN 2). They subsequently formed the backbone of GSM type approval. However, when GSM was extended to add packet radio, GPRS, prose versions of test cases were duly developed by the appropriate ETSI forum, but there was no organized effort to provide reference test implementations. It is commonly recognized that this contributed towards the delays and problems in the early deployment of GPRS. As a result of these experiences, it was decided from an early point that 3GPP should be responsible for the development of reference test implementations for signalling tests, and this effort was funded mainly by 3GPP.

5.2 Specifications

The UMTS system is defined by more than 2200 3G specifications: some inherited directly from GSM, some inherited and then substantially extended and a few which are unique to the third-generation standard. The specification process is a continuous one, with new features continually being added to the standard as new services or capabilities are added. To cope with this flux, 3GPP has evolved a rather complicated process of releases and stages.

5.2.1 Releases

With any communications system of this magnitude, the requirements it must meet are large and likely to grow further over time. From the outset of UMTS, there was significant pressure to get at least a basic system commercially deployed as quickly as possible. Some of this pressure came from Japan, where the second-generation networks were largely running out of capacity, and some came from Europe, where large amounts of capital had been invested in operating licenses. To allow a basic system to be specified, but still allow enough flexibility for the system to grow in capability in the future, a system of releases was introduced. Each release builds on the capabilities of the previous one.

The first official complete release of the 3GPP specifications is officially known as 'Release 99', but is sometimes also referred to as 'Release 3'. The term 'Release 99' came from the initial target of the standards forum to complete the specifications by the end of 1999. A large part of the work was completed within this timescale, and the specifications were officially frozen in June 1999. However, at this point there were still enough significant problems with the specifications that commercial operation was not realistic. The freeze prevented the addition of any more features or capabilities to the specifications for this release and meant that only changes to fix critical errors were allowed. However, the specifications did not achieve a reasonable level of stability until June 2001, which formed the base version for many early network operations.

Following Release 99, 3GPP moved away from using year numbers to denote releases and switched to using the first number of the version triplet. Hence, the next release was 'Release 4', which for a while was called 'Release 00' for the year 2000, and the release after that was 'Release 5' and so on. Each release encompasses many varied features, and while there may be a few headline items, such as the introduction of high-speed downlink packet access (HSDPA) in Release 5, generally the releases are a collection of features addressing a number of different market requirements.

5.2.2 Work Items

To distribute the work effectively among the various working groups, there is a hierarchical decomposition of the work required to realize each feature within a release. A feature is subdivided into a number of building blocks, where each building block is intended to relate either to a single component within the system or to an interface, a protocol or protocol layer. Building blocks are further subdivided into work tasks. The scope of a work task should be solely within a single TSG working group. In fact, due to the logical structure within the 3GPP TSGs, often even the work within a building block is limited to a single working group. Features, building blocks and work tasks are all collectively referred to as work items. The main purpose of this hierarchy is to distribute the work relating to a feature, whilst keeping the responsibilities for each task clear, so naturally the decomposition stops at the point where this is the case. Depending on its complexity, not every feature has more than one building block, and in many cases the building blocks are not broken down into work tasks.

A good way to visualize this in practice is to look at an example, such as HSDPA. This is itself a feature, which was decomposed into a number of logical entities: the air interface work (layer 1 – L1), the MAC and RLC enhancements (L2), the RRC extensions (L3), modifications to the transport interfaces within the network (I_{ub}, I_{ur} and I_u), the radio performance requirements and, of course, conformance testing of the feature. These make up the building blocks for HSDPA. As each building block happens to fall neatly into the domain of a single RAN working group, there is no further decomposition (Figure 5.2).

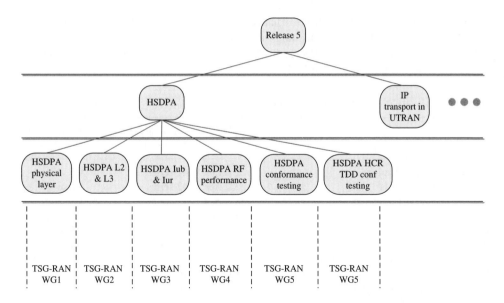

Figure 5.2   Work item structure for Release 5, HSDPA

5.2.3 Stages

This then leads to the question of how the features within a release are decided on. To perform a meaningful decomposition, 3GPP must have some idea as to what is required to realize a feature. They must also have some idea as to how long it will take to complete the standardization. For smaller tasks, this may be quite straightforward to estimate, but for large features, a degree of preparatory work is needed. Typically, for any significant new feature, this starts with a feasibility study ending with the release of a technical report. This report will usually set out, at quite a high level, some options for the new feature and may recommend a way forwards. For features within the mid range, this is often all that is required. The technical requirements for the feature are either established in the technical report or left to the individual working groups to establish. Again, following the example of HSDPA, the technical report was carried out as part of the Release 4 work and published in TR 25.950 version 4.0.1. For more significant features, a staged process is followed.

Stage 1 is essentially to gather the overall system requirements. It is officially defined as the overall service description from the user's standpoint. In stage 2, the architecture of the network realization of the feature is defined, with the aim of mapping the requirements from stage 1 into the blocks, entities, interfaces and protocols needed. Stage 3 is effectively the detailed design, where specifications are drafted for those parts of each component that need to be standardized. This is an important point. Unlike the specification that a company might create for its own product, where the full design may need to be documented, a standards forum intends only to specify what is needed to ensure independent implementations can operate together. Therefore, the specifications tend to focus mainly on what should be done, rather than how it should be done, and where possible they try to leave as much freedom as possible for the implementer. This three-stage approach actually comes from an ITU-T recommendation, I.130. Although not usually referred to as such, the development of the conformance test specifications constitutes stage 4.

This staged approach is also reflected in the organization of the 3GPP specifications into numbered series, as shown in Table 5.1.

Table 5.1 Organization of 3GPP specifications into series (reproduced by permission of ETSI)

	UMTS	GSM only (Release 4 onwards)	GSM only (up to Release 4)
Requirements specifications	21 series	41 series	01 series
Stage 1: Service aspects	22 series	42 series	02 series
Stage 2: Technical realization	23 series	43 series	03 series
Stage 3: UE to CN signalling	24 series	44 series	04 series
Stage 3: Radio access	25 series	45 series	05 series
Stage 3: Audio and video coding	26 series	46 series	06 series
Stage 3: Terminal data modem services	27 series		(07 series)[a]
Stage 3: Radio subsystem to CN signalling	28 series	48 series	08 series
Stage 3: CN internal signalling	29 series	49 series	09 series
Stage 3: Subscriber Interface Module	31 series	51 series[b]	11 series[b]
Charging and OAM&P (Operations, Administration, Maintenance and Provisioning)	32 series	52 series	12 series
Network security	33 series		
UE-related conformance test specifications	34 series	51 series[b]	11 series[b]
Security algorithms	35 series	55 series	

[a] These relate to functionality within the UE which is not directly attributable to a specific radio technology; therefore the old GSM 07 series were updated to 27 series, and 47 series was never created.

[b] This is a legacy problem. The old GSM conformance specification was numbered 11.10, and it was felt less confusing to keep with the convention of adding 40 to produce the new 3GPP numbers rather than renumber it and create a 54 series.

5.3 Conformance Specifications

The intention was for the test specifications to fall within the 34 series, and originally a TSG for issues relating to the terminal (TSG-T) was formed and its working group 1 (known as TSG-T1) was given responsibility to develop specifications for conformance testing. However, for certain areas, specifically the Universal Subscriber Identity Module (USIM) and audio areas, the test specifications are developed by the same group that develops the core specifications. This is accepted as a practical way forwards as knowledge in these areas is quite specialized and not usually present in the main conformance test working group. This means that TSG-T1 concentrated exclusively on RF aspects, signalling and lower layer protocols and electromagnetic compliance. This is reflected in the organization of the test specifications, which are listed in Figure 5.3, together with the 3GPP specification group responsible. In late 2004 and early 2005, 3GPP was reorganized and the terminals TSG restructured with the working groups all moving to other TSGs. The conformance specification group was moved into the RAN TSG, where it became working group 5

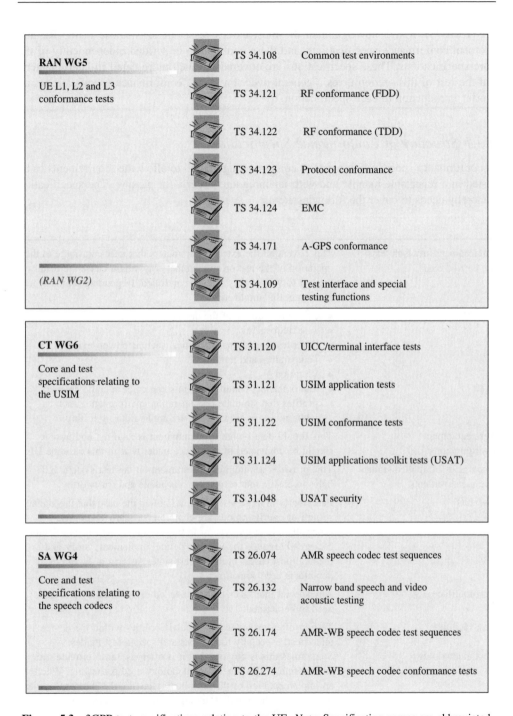

RAN WG5	TS 34.108	Common test environments
UE L1, L2 and L3 conformance tests	TS 34.121	RF conformance (FDD)
	TS 34.122	RF conformance (TDD)
	TS 34.123	Protocol conformance
	TS 34.124	EMC
	TS 34.171	A-GPS conformance
(RAN WG2)	TS 34.109	Test interface and special testing functions
CT WG6	TS 31.120	UICC/terminal interface tests
Core and test specifications relating to the USIM	TS 31.121	USIM application tests
	TS 31.122	USIM conformance tests
	TS 31.124	USIM applications toolkit tests (USAT)
	TS 31.048	USAT security
SA WG4	TS 26.074	AMR speech codec test sequences
Core and test specifications relating to the speech codecs	TS 26.132	Narrow band speech and video acoustic testing
	TS 26.174	AMR-WB speech codec test sequences
	TS 26.274	AMR-WB speech codec conformance tests

Figure 5.3   3GPP test specifications relating to the UE. *Note*: Specification names are abbreviated.

(TSG-RAN5). RAN5 now maintain the most extensive set of tests: those in the RF and protocol conformance specifications, and these continue to be written independently of the core specifications. These specifications are covered in much more detail in later chapters, but the rest of this section gives a general overview of the conformance specifications and a brief introduction to each specification.

5.3.1 Structure of Conformance Specifications

A conformance specification needs to contain enough detail to allow the requirements to be tested in a repeatable manner and with unambiguous criteria for passing. The specification generally needs to cover the following items:

Environment in which tests are carried out	This covers all the external parameters that may contribute to the operation of the test or, put another way, might cause the result of the test to vary if they are not controlled. Depending on the type of testing, this might include: • Operating voltage of the device under test • Test frequencies • Other test signal characteristics, such as power level • Temperature and humidity • Vibration • Radio channel conditions – in some cases specific channel profiles that simulate real operating environments, such as white noise, or slow fading are applied during testing
Test equipment configurations	This should detail what test equipment is required and how it should be connected to the device under test; in this case the UE.
Accuracies of applied signals and measurements	This provides a minimum requirement on the test system and helps to ensure that results are repeatable and trustworthy.
Defaults	Particularly in large test suites, it is often the case that the signals, stimuli or configurations are common for many test cases, especially where the type of testing is similar. The use of defaults can greatly reduce the work to define, implement, verify and maintain test cases. In the largest suites, a form of hierarchy of defaults is used sometimes.
Test interfaces	Any special interfaces that are needed or assumed by test cases need to be defined.
Test facilities	Testing often makes use of special facilities within the device under test, especially loop-backs and special test modes.
Test methodology	Sometimes this is obvious, but it is often useful to provide some explanation of the general methodology used for testing. Where an implementation of the tests is provided by the standards group, this serves to help understanding of how the implementation was done, and where it is not provided this acts as a guide to implementers.

(continued)

Implicit testing statement	In larger test suites, many of the basic or underlying functions are heavily exercised during normal test operation. An example of this is the normal operation of ciphering. Typically it will be enabled and operating during signalling tests, and any problems with the normal operation will surface somewhere during the hundreds of signalling tests run. It is often very useful for the standards group to state their assumptions as to what functionality will be implicitly tested.
Test case descriptions	This is, of course, the meat of the conformance specification and many different ways are used to describe test cases. These are organized into logical groupings.
Applicability	Testing anything but the simplest of most fundamental functions will bring in an element of optionality. The test case may only apply to certain classes or types of device, or may only be appropriate if certain features are present. In some cases, the applicability of a test case may depend on quite a complicated interrelationship between different features.
Test case implementations	In some cases, the standards forum chooses to provide test cases in a form that can almost directly be run on commercial test equipment. The standards body usually has to take care not to show favouritism to individual suppliers, so the test cases are abstracted from specific test systems, and the equipment manufacturers are given a well-defined interface to adapt their equipment to. These implementations are therefore referred to as abstract test suites (ATSs), and generally use a standardized test language or notation for which a variety of commercial tools are available. For ETSI and 3GPP, the test language of choice is either TTCN 2 for older test suites or TTCN 3 for more recent ones. These languages are discussed in more detail in the Appendix and Section 14.3.1, respectively.

Where the scope of the conformance specification is relatively small, these can be contained within a single document. However, for larger areas this approach can lead to large specifications which are difficult to navigate, maintain and understand. These items can be spread over a number of specifications, or the specification can be divided into parts, or sometimes there is a combination of both.

5.4 RAN5 Specifications

TSG-RAN5 is responsible for the major part of UE equipment testing and its work is split into two areas: the radio and the signalling protocols. RAN5 is organized into two subworking groups (SWGs), one for each area. In addition to these conformance test areas, there are certain areas of commonality which lead to the creation of two auxiliary specifications, TS 34.108 and TS 34.109. As they do not fall into any of the specific test areas covered in detail in later chapters, they are described in a little more detail here.

5.4.1 TS 34.108

> Universal Mobile Telecommunications System (UMTS);
> Common test environments for user equipment (UE);
> Conformance testing

The RF and signalling areas require very different sets of tests, but there are still areas of significant commonality, such as the network environment under which the tests are conducted. To simplify maintenance, and avoid having duplicated information, the common parts related to all types of testing were placed in a single specification, TS34.108, titled 'Common Environments for User Equipment (UE) Conformance Testing'. The document contains a number of diverse sections which provide much of the background information needed to understand the conformance test specifications; in particular it covers the following areas:

Section	Purpose
Common requirements of test equipment	The aim is to define a minimum set of requirements that conformance test systems should meet. They are very high level, and of course the full requirements are set by the test cases the equipment is intended to run. This section does include some useful information on general equipment accuracies and overall capabilities, such as the minimum number of each type of channel needed.
Reference test conditions	This details a number of general test parameters which are widely used in tests, such as the frequencies at which testing should be carried out within each frequency band, and the allocation of orthogonal variable spreading factor (OSVF) channel codes to physical channels. Historically, this section started as a broader description of the test environments, but parts that were specific to certain types of testing, such as the descriptions of the various radio channel simulations which are quite specific to RF testing, were later moved to the relevant test specifications, and replaced by pointers in this section.
Reference system configurations	This is one of the key sections of this document. It describes in detail the simulated network conditions, parameters and settings used during conformance testing. In practice, every network is set up differently depending on many factors, such as geographical considerations, combinations of infrastructure equipment and the service strategy of the operator.

(continued)

Section	Purpose
	The conformance test system has to 'fool' the UE into believing it is operating in a real network so the system configuration needs to look realistic without actually copying one specific network set-up. In fact, the reference configurations were written before fully compliant UMTS networks started to operate, and in some areas they set a good starting point for operators and equipment manufacturers to follow.
	There are also some general operating settings where the network operator has a great deal of flexibility, and in many cases these relate to advanced features some of which may not be used in real networks for some time. In the conformance 'generic' network, these types of features need to be enabled where possible to make sure the UEs are capable of operating successfully when these features are eventually switched on. Examples of this are transmit diversity, compressed mode settings and power control settings. Many of these network settings are broadcast in the system information, and this is included in the default message contents section described below.
• Simulated network environments	Conformance tests are not run in a completely emulated network environment. Each test has a specific execution life cycle where an environment is created, the test run and a verdict reached. The environment is an abstraction of real life, with only those parts of the network necessary to achieve the purposes of the test included. Of course, it does not make sense to create a different environment for each test case, and in fact many tests have very similar environment requirements. Therefore a number of different environments for different areas of testing are identified in this section. In line with the general strategy, these form the defaults, and typically if a test case has very specific requirements, only the differences from the default need to be specified. The system information is defined in a hierarchical way, with a basic set for all the various environments, together with a set of deltas for adding further cells and for the different network configurations.
• Reference radio bearer configurations used in radio bearer interoperability testing	This is a very important subsection, with implications much wider than conformance testing. A significant thread of 3GPP's strategy was to create a 'toolbox' of functionality which would provide a lot of flexibility for network operators to provide differentiated services. In GSM the physical bearer was relatively fixed, with very few parameters that could be varied. In UMTS there are a very large number of variables. This provides enormous flexibility, but potentially presents a huge challenge for testing. The number of possible radio bearer configurations is very large, and it would simply not be practical to test all of them. Furthermore, the number of variables and

(continued overleaf)

(continued)

Section	Purpose
	their complex interactions makes it unpractical even to perform boundary testing. The solution adopted by 3GPP to this was to define some specific radio bearer configurations that represented realistic traffic bearers. These bearers have now become a reference for the industry as they form a set that all the UEs are tested against. Operators know that while they stay on, or close to, these configurations, they can guarantee operation with any conformant UE. The bearers are classified by traffic class and rate and are used in testing in two ways: the first is that a suitable configuration is selected from the list, according to test purpose and UE capabilities, when a bearer needs to be established as part of a signalling test; and second, each bearer configuration is rigorously tested to ensure it is carrying traffic correctly.
• Common radio bearer configurations for other test purposes	The main practical drawback with the reference radio bearer approach arises from the large number of UMTS networks. Radio bearer configurations are compromises which trade-off many variables, such as network capacity, robustness and speed, against each other. No two networks are identical, and operators may have different strategic aims and commercial drivers. Therefore the balance of trade-offs can differ quite widely and without some care, the number of references could have easily grown to unmanageable proportions. The standards group had to make sure that bearer configurations in the reference set represented a democratic majority of networks, and that duplication with 'fairly similar' configurations was avoided. Strict rules were brought in to control the introduction of radio bearers to the reference set. However, in certain cases very specialized configurations are required to test some features either because the feature is tested in a special environment to permit greater test coverage or because it is provided for future deployment and at the time of approval of that version of the specification there is no industry consensus on a real-life configuration. Examples of the former are testing of the MAC and RLC layers, where special configurations are needed to exercise these layers and create the necessary boundary and failure conditions.
Generic setup procedures	Test cases begin with a preamble that brings the UE into a state where the specific test can be carried out. For example, if the purpose of the test is to check the behaviour of the UE when the radio link is lost during the set up of a call, the preamble must first get the UE into a state where the call can be started. Originally (Release 99), the procedures themselves were organized into four subsections: basic generic procedures, generic setup procedures, test procedures for RF test and common generic procedures for AS testing.

(continued)

Section	Purpose
	There are only two basic generic procedures: the setting up of an RRC connection, and the establishment of a radio bearer. These are essentially macros that are used in a variety of test cases, including the procedures that follow in the later subsections. The second and third subsections cover the initial states used in NAS and RF tests, respectively. The last subsection is the most complex, covering RRC and other access stratum testing, where there are many states that tests can start from, and often several ways to reach a particular state. In TS 34.108 this is handled in quite an elegant way. First a map was created of all the relevant stable states that the UE could pass into or through. The map is organized as a dependency tree, with each state linked upwards to the preceding state or states, and downwards to the succeeding ones. Each link then has a procedure associated with it, labelled with 'Pn', where, in Release 7.0.0, n is a number from 1 to 26. The procedures are designed in a modular way, so to transition from Power Off (state 1) to CS-CELL dedicated channel (DCH) Initial (state 3), procedures P1 and P3 are executed sequentially. Since Release 99, two further subsections have been added: test procedures for Assisted Global Positioning System (A-GPS) performance requirements testing, and test procedures for Multimedia Broadcast and Multicast Service (MBMS) testing.
Test USIM parameters	The USIM holds some key information which characterizes and controls the behaviour of the UE, such as the identity of the home network. This information needs to be defined to predict UE behaviour, and in some cases the USIM content actually forms a key part of the test case. This presents a real challenge for automating the operation of conformance tests. Having a test setup that allows continual execution of all the test cases without operator intervention is somewhat defeated if the test flow has to continually be interrupted to open up the UE and change the USIM card. At the time of writing, there are around nine different USIMs required to cover the current Release 99 test cases used for certification. As well as network information, the USIM also holds parameters associated with authentication, ciphering and integrity, and it holds the key generation and checking algorithm. In real networks, this algorithm is proprietary to the network operator, and is strictly protected. For conformance testing, a special algorithm is used. Test USIMs already programmed with the algorithm can be purchased from a number of suppliers. The algorithm is detailed in this subsection and has been designed to allow full testing of security features without giving away confidential network details.

(continued overleaf)

(continued)

Section	Purpose
Default message contents	This section contains a set of message contents for each RRC message used in signalling tests. The specifications operate a hierarchical way of specifying the message content, and the message definitions in this section form the base or default level. Where specific message content is required in an individual test case, only the deltas from this section are specified. Only the RRC messages are treated in this way, and this is mainly for historical reasons. The NAS test cases were originally transferred from the GSM test specification (GSM 11.10 as was). As the SMG had not taken a hierarchical approach, it would have created considerable extra work to go through all the NAS tests and rework the message contents. In addition, the same NAS signalling is used for both GSM and UMTS, and consequently there is considerable commonality between the test specifications. There has been, and still is, a lot of cooperation between the two standardization groups, with changes and fixes identified in one group replicated by the other. This cooperation was made much simpler by keeping the two approaches the same.
A-GPS GPS scenarios and assistance data	A-GPS is a special case compared to most conformance testing as it requires the use of a second simulator to generate simulated satellite signals. In order for the results of any test to be predictable, and for any information sent by the network for assistance purposes to be meaningful, all of the environmental parameters used to calculate position need to be defined. This section provides definitions for environments and corresponding message contents for both RF (performance) and signalling tests. For the purposes of these test cases, the GPS simulators are loaded with a set of parameters that represent the starting point of the environment, and are then run. The simulator will then calculate the movements of the satellites and vary the signals accordingly. Conveniently, a file containing the configuration data for setting up the GPS simulator is attached to the electronic distribution of TS 34.108.
MBMS configurations for signalling test	This section contains some scheduling configurations relating to the transmission of the control information broadcast in MBMS signalling tests. At the time of writing, this section is still under development.

5.4.2 Hierarchy of Test Parameters

Using defaults is an efficient way of defining test cases as it reduces the amount of work needed to define each individual test case and reduces the amount of maintenance work. However, with a large body of tests it is difficult to create a set of defaults that apply widely enough, and large numbers of tests can require parameters at variance with the defaults.

The solution to this problem is to establish a hierarchy, and this is the approach used by RAN5. The defaults in TS 34.108 form the base level of the hierarchy. The next level of the hierarchy comes from the definitions at the start of a new clause, or section in the test specification itself. Where tests are grouped into logical subsections, any common definition at the start of a subsection forms the next step in the hierarchy, and finally any specific message contents included in the test case description itself form the top of the hierarchy. Parameters defined higher up the hierarchy take precedence over parameters defined lower down. Hence, if a message information element (IE) is given a value in the specific message content of the test case, this value will override the value given in the section introduction, which in turn will override the value given in TS 34.108.

5.4.3 TS 34.109

> Universal Mobile Telecommunications System (UMTS);
> Terminal logical test interface;
> Special conformance testing functions

A certain amount of testing can be performed by connecting a UE to a simulated network and sending it a predetermined set of signalling messages. However, for some layers or functions this is not sufficient. Usually two additional things are needed: some method of triggering the UE to initiate actions itself, and some method to cause the UE to transmit certain data content back to the simulator. There is a general goal when developing conformance tests to try to minimize any overhead on the UE associated with, or specific to, testing, so providing these functions requires some care. It is these 'special conformance testing functions' that form the subject of this specification. This specification has an important difference from any other test specification. The others define tests that assure the requirements set out in other 'core' specifications but do not in themselves place any requirements on the UE. TS 34.109, in contrast, contains specific requirements on the UE that do not derive from any of the core specifications. It was therefore developed originally by the conformance test TSG (T1), but with an understanding that once the test aspects had been covered within this specification it should be passed to RAN2 for approval and future maintenance. This decision had a useful consequence. All of the protocol-related implementation requirements for the UE are under the control of a single specification group, and arguably, this has led to a more consistent implementation of the test-related requirements than happened under GSM.

Section	Purpose
Test control (TC) protocol procedures and test loop operation (including message definitions and contents)	The principal way of getting the UE to send back specific data to the simulator is through the use of a test loop. This allows the simulator to transmit content to the UE using a downlink channel. The UE will then retransmit the content back to the simulator following whatever protocol rules are configured in

(continued overleaf)

(continued)

Section	Purpose
	the layers below the loop-back point. These test loops are very important and are covered in more detail in Chapter 8. Associated with the definitions of the test loops are some associated signalling messages that enable or disable the operation of the loops.
Electrical man–machine interface (EMMI)	There was a great deal of interest within the industry to be able to automate operation of the conformance tests. Just for Release 99, the protocol conformance tests take 1 to 2 weeks to run manually. The use of test loops helps to enable the test system to provide all the necessary data for testing, but still many tests require events to be triggered from the UE. In GSM an attempt was made to define a standardized interface which the test system could connect to and control the UE in the same way that a human operator might. This was called the electrical man–machine interface (EMMI), and is detailed in GSM 11.10-1 (now TS 51.010-1). Unfortunately this never quite worked out. Manufacturers were sometimes reluctant to spend too much effort on building an interface for test purposes only, and in hindsight the specification was a little complicated for benefits it provided. However, with UMTS the interest had not gone away, and a more flexible and simple approach was developed, based on the use of standard AT modem commands, which are quite widely supported by standard terminals.
Universal integrated circuit card (UICC)/mobile equipment (ME) test interface	In common with most functions within the UE, the interface with the USIM card is standardized so that cards from different manufacturers can work in any terminal, and this interface is subject to a suite of conformance tests. The interface itself defined in TS 31.101, and this section just contains a pointer to it.

5.4.4 TS 34.121

> Universal Mobile Telecommunications System (UMTS);
> User equipment (UE) conformance specification;
> Radio transmission and reception (FDD)

As the title suggests, this is the specification that describes the conformance tests for the radio transmitter, receiver and baseband sections of the UE. The core requirements behind these tests are defined within RAN4, which is responsible for defining the radio performance requirements of the system (both UE and network side). The RAN4 specifications cover a number of areas, but the main ones relevant to TS 34.121 are given in their context in the following paragraphs.

TS 34.121 is split into two parts; these are denoted by a '-1' or a '-2' following the specification number. As discussed earlier in this chapter, for more complicated features, when defining the conformance tests the standards body also has to indicate which class or type of UE the test applies to. This is called the applicability statement. For reasonably large conformance specifications including this information in the test definition itself is undesirable, as it becomes very difficult to work out, for a given model of handset, what test cases need to be run to prove conformance. Separating the applicability information into another specification provides a much more manageable approach. In this case, TS 34.121-1 contains the test cases and TS 34.121-2 contains the applicability. Incidentally, one consequential benefit of this is that it allows much better version management of the large test specifications.

The most important sections of TS 34.121-1 are as follows:

Section	Purpose
Frequency bands and channel assignments; reference conditions	These sections are mainly informative and present information, such as the frequency band definitions, which are defined in other core specifications. The reference conditions are mainly held in TS 34.108 (see above).
Transmitter characteristics	This is referenced to TS 25.101 section 6. Each specification, for example, transmit power, frequency error and so on, has an associated test case.
Receiver characteristics	This is referenced to TS 25.101 section 7, and again the limits and tolerances allowed for each characteristic are set in the core specification and tested in this section.
Performance requirements	This is referenced to TS 25.101 section 8, and generally the tests measure block error rate (BLER) in a variety of link configurations and channel conditions.
Requirements for support of radio resource management (RRM)	RRM is covered in more detail in Chapter 12, but represents a blend of both RF and signalling testing. The majority of tests fall into two categories: those that check timing and those that check measurement accuracies. There are a number of requirements on the UE to perform actions within a certain time limit, where its ability to meet the requirement depends on being able to receive, demodulate and interpret signals correctly. The actions the UE takes might often be interpreted as signalling, but the decisions behind those actions depend on the performance

(continued overleaf)

(continued)

Section	Purpose
	of its radio subsystem. There are also many measurements made by the UE that trigger signalling reports, and the success of the signalling depends on the ability of the UE to measure within the accuracy required by the specifications. In general these tests are referenced to TS 25.133, although some of the timing and accuracy requirements come from specifications such as TS 45.008 (GSM 05.08), TS 25.214, TS 25.215 and TS 25.303.
Performance requirement for HSDPA, for E-DCH and for MBMS	These three sections have been added for the Release 5 and later physical channels, and their requirements come from TS 25.101 sections 9, 10 and 11, respectively.
Connection diagrams	This annex shows a series of test equipment connection diagrams that are used for the various tests. In reality, RF test equipment will achieve the various set-ups by switching equipment in and out. However, some of the RRM tests configurations require simulation of several neighbour cells, and will use a different or extended test system.

5.4.5 TS 34.122

> Universal Mobile Telecommunications System (UMTS);
> Terminal conformance specification;
> Radio transmission and reception (TDD)

TS 34.122 is the equivalent specification to TS 34.121, but covers the TDD mode of operation. It is organized in the same way with the exception that it has not been broken into parts at this time.

5.4.6 TS 34.123

> Universal Mobile Telecommunications System (UMTS);
> User equipment (UE) conformance specification;
> Part 1: Protocol conformance specification

This is the largest of the conformance specifications, stretching to over 4500 pages and containing over 1400 tests. It is divided into three parts:

TS 34.123-1	This contains the conformance test case descriptions and is covered in more detail below and in Chapters 9 and 10.
TS 34.123-2	Titled implementation conformance statement (ICS) specification, this contains the applicability of each test case. This can be quite difficult to work out as for any particular test case it can depend on a number of different features or capabilities being present, or absent, in the UE. The part 2 specification contains a set of tables that list all the options possible for a UE. The device manufacturer can go through these tables ticking the options that apply to his device. This creates an ICS, essentially a statement of what has been implemented. Each line in each table has a unique identification, and these are then fed into the conditional statements to decide if a test case applies or not. This whole process is easily automated, and support for this methodology is already an intrinsic part of the TTCN 2 language and tools.
TS 34.123-3	Titled ATSs, this comes in two parts. The most important part is an electronic distribution of the source code for the tests that have been implemented in TTCN 2. This is a series of files of type '.mp', which are ASCII text, but are not easily understandable to a reader. However, there are a number of commercial tools that will take these files and render them into a readable format. There are also tools that will generate executable test cases from them. The second part is a document that accompanies the ATSs and provides a wealth of information about the test methodologies and about the architectures and interfaces used in the test suites.

In outline, the key sections of TS 34.123-1 are structured as follows:

Section	Purpose
Idle-mode operations	This section tests the NAS-related idle mode tasks of PLMN selection and roaming, and the AS idle mode tasks of cell selection and reselection. These tests are generally the most complicated in terms of the test environment, as they require the presence of many cells in the simulation. The general format of these tests also tends to be a little different to the later sections, as the tests are generally about setting up certain RF conditions (e.g. power levels) on the cells and waiting for the UE to perform some action. The signalling is relatively simple, and usually not defined in detail. The NAS requirements trace to: • TS 23.122 (NAS functions related to the MS in idle mode), which is actually a stage 2 specification, but the only place where the system level requirements come together, and in some cases, related to this: • TS 24.008 (NAS signalling), where some of the actions the UE must take relating to network signalling are specified • TS 25.304 (UE procedures in idle mode and procedures for cell reselection in connected mode), which covers the AS functionality • GSM 03.22 (functions related to MS in idle mode and group receive mode), which covers some of the functionality related to GSM interworking.

(continued overleaf)

(continued)

Section	Purpose
Layer 2	This covers testing of the layer 2 core specifications for the MAC, RLC, PDCP and broadcast/multicast control (BMC). Testing of layer 2 in an integrated product is always a challenge as there are usually no externally exposed direct interfaces to these layers. Originally, in GSM phase 2 it was only implicitly tested in that it was used to carry the signalling, but this only allows limited testing of recovery from failure scenarios. In UMTS enough flexibility existed in the normal operating modes of the upper layers, which together with some test loops allowed more innovative approaches to fully exercising these entities. This is covered in more detail in Chapter 9.
Radio resource control (RRC)	Physically, this is the largest section, running to almost 2000 pages in total, and this indicates the complexity of the 3GPP RRC protocol. The first two subsections, RRC connection control and radio bearer control, are fairly straightforward. The next two sections, covering RRC connection mobility and measurement procedures, deserve further explanation. The mobility procedures have some overlap with the idle mode procedures, in that both sections are testing cell reselections. However, this section contains the connected mode reselections, and the emphasis is more on the signalling. This subsection also includes state transitions between all of the connected mode RRC states using the Cell Update procedure, which are not necessarily mobility related. The lines are a further blurred because hard handover, which would normally be thought of as a mobility procedure, is covered under radio bearer control. The reasons for this are due to the flexibility of the RRC. Many of the messages or procedures can be used to perform a variety of different actions. For example, hard handover is just one case of the physical channel reconfiguration procedure (among others). It still makes logical sense to try to organize the tests by procedure, but sometimes the functional boundaries do not fit very well. The measurement procedures also have some commonality with the detailed testing of measurement carried out in the RRM section of TS 34.121 (FDD) and TS 34.122 (TDD). The main difference is that in this section the RF conditions are not so precise, and no attempt is made to check the accuracy of the UE's measurements. The tests verify the signalling procedures as specified in TS 25.331. However, there clearly is some overlap. It is not possible to verify the measurement accuracies in TS 34.121/122 without executing at least some of the signalling. To avoid duplication, some rationalization of the procedures was done between the RF and signalling SWGs, and signalling that is implicitly tested in the RF tests is not necessarily retested in this section.
Elementary procedures of mobility management (MM);	These four sections test the main NAS protocols, and were inherited from GSM. These tests were copied from the Release 97 version of the GSM test specification and updated initially when the NAS protocols were extended for UMTS. Since then the NAS has been further extended, for both GSM

(continued)

Section	Purpose
Circuit switched call control; Session management (SM) procedures; Elementary procedures for packet switched MM	and 3G, and so these updates have to be applied back into both test specifications in a synchronized way. This has been done by a close cooperation between both conformance specification groups.
General tests	This section was intended to be a 'catch-all', but through various early reorganizations of the document now only contains the tests for emergency calls. From a legislative perspective, these are some of the most important tests in the document, as some regions have legal requirements on networks to provide access to emergency calling.
Interoperability radio bearer tests	This section relates directly to the reference radio bearers defined in TS 34.108 (see above). Each reference radio bearer has a specific test case, but the test cases follow a common format and test method, the variations being the radio bearer parameters used. One of the rules that TSG-RAN5 introduced to ensure that reference radio bearers had enough industry support, was that in order to gain acceptance for adding the radio bearer into TS 34.108, the supporting companies had to commit to add a test case to this section within 6 months. TSG-RAN5 also placed a requirement that the supporting companies had to commit resource to provide a TTCN test case in TS 34.123-3 within a further 6 months.
Supplementary services	So far no 3G supplementary services tests have been defined. However, the underlying SS protocol is relatively straightforward and is tested under GSM.
Short Message Service (SMS)	The SMS tests were also inherited directly from GSM.
Specific features	The tests in this area are intended for services or features that are in some way related to other protocols, but are discrete enough to warrant their own subsection. Currently, two such features are defined: • Autocalling restrictions • Location-based services (LCS) Both of these potentially have legal requirements in some regions. Autocalling restrictions relate to the rules surrounding auto-dialling or auto-callback features, particularly of the modem component, to prevent repeat calls becoming a nuisance, and LCS refers to the provision of services based on the network being able to determine the precise location of the handset, usually using GPS. In this case though, the tests are actually of the signalling between the handset and the network to determine position. There are no tests of actual services.

(continued overleaf)

(continued)

Section	Purpose
Multilayer functional tests	In theory this section should cater for vertical tests, that is, testing paths through all the protocol layers such as is done in end-to-end testing. However, currently this section is used to hold radio bearer tests for configurations that cannot make it into the reference radio bearer list as they do not have wide enough support, but are clearly needed to test certain parts of the standard. Currently, the section contains some radio bearer tests for TDD mode.

As a footnote, one of the annexes is titled 'Default RRC message contents', but during the development of the Release 99 specification these were moved to TS 34.108 to keep all of the default message content in one place. This also created a clearer hierarchy of defaults as discussed earlier in this section.

5.4.7 TS 34.124

> Universal Mobile Telecommunications System (UMTS);
> Electromagnetic compatibility (EMC) requirements
> for mobile terminals and ancillary equipment

This specification covers the tests and test environments to check that the UE, together with any associated add-ons, does not generate excessive electromagnetic interference, and that it is not overly susceptible to external interference. In fact, this started life within TSG-T1, along with the other conformance specifications. Unlike the other areas, though, the requirements for EMC compliance are not specific to UMTS devices and are generally regulated by government bodies. As UMTS networks operate globally, the requirements were derived from internationally harmonized specifications, mainly from the ITU, the International Electrotechnical Commission (IEC) or its International Special Committee on Radio Interference (CISPR), or from the International Standards Organisation (ISO). Because these are not 3GPP core specifications, they need to be introduced as core requirements on UEs, and in line with the policy that the conformance test working group should not introduce core requirements itself, the specification was subsequently transferred to TSG-RAN4, which now maintains it.

5.4.8 TS 34.171

> Terminal conformance specification;
> Assisted Global Positioning System (A-GPS);
> Frequency Division Duplex (FDD)

Related to the LCS signalling tests in TS 34.123, this specification tests the actual performance of the GPS receiver within the UE. Mostly, this is in terms of achieving a positional accuracy within an allowed time limit. The performance requirements for A-GPS come from TS 25.171 'Requirements for support of Assisted Global Positioning System (A-GPS); Frequency Division Duplex (FDD)'. One point to note is that A-GPS can be implemented in the UE at two levels. The UE can include a complete GPS system, or it can be a partial implementation that passes measurements back to a location server which makes the positional calculations, provides the maps and mapping support and so on. In the latter case, to determine a pass or fail, these calculations have to be done on the test results.

5.5 Other Conformance Specifications

5.5.1 Universal Subscriber Interface Module

There are a number of test specifications focussed around the USIM functionality (Table 5.2), but these fall into two categories: tests of functionality on the USIM itself, and tests of functionality on the terminal. The tests on the USIM are outside the scope of this book, but some of the terminal related tests are quite important as they form part of the various certification criteria (CC).

Table 5.2 Summary of Universal Subscriber Interface Module (USIM)-related test specifications

Specification		USIM	Terminal
TS 31.048	Security mechanisms for the (U)SIM application toolkit; Test specification	✓	
TS 31.120	UICC-terminal interface; Physical, electrical and logical test specification[a]		✓
TS 31.121	UICC-terminal interface; Universal Subscriber Identity Module (USIM) application test specification		✓
TS 31.122	Universal Subscriber Identity Module (USIM) conformance test specification	✓	
TS 31.124	Mobile Equipment (ME) conformance test specification; Universal Subscriber Interface Module Application Toolkit (USAT) conformance test specification.		✓
TS 31.213	Test specification for subscriber (U)SIM; Application Programming Interface (API) for Java Card™.	✓	
TS 34.131	Test Specification for C-language binding to (Universal) Subscriber Interface Module ((U)SIM) Application Programming Interface (API).	✓	

[a] This document does not actually contain any tests, but is a pointer to another ETSI specification TS 102 230 'Smart cards; UICC-Terminal interface; Physical, electrical and logical test specification', which is not itself a 3GPP document.

5.5.2 Codec Testing

Testing of the audio codec (coder/decoder) in the UE is mainly covered in two closely related specifications: TS 26.131 and TS 26.132 (Table 5.3). The former sets out the performance requirements and the latter describes the test cases. These codec tests can form part of the

Table 5.3 Audio and video test specifications

	Specification
TS 26.131	Terminal acoustic characteristics for telephony; requirements
TS 26.132	Speech and video telephony terminal acoustic test specification
TS 26.074	AMR speech Codec; test sequences
TS 26.174	Speech codec speech processing functions; Adaptive Multirate – Wideband (AMR-WB) speech codec test sequences

CC, and some of the tests themselves cover issues such as maximum loudness at the UE's earpiece, and may have a relationship to product safety regulation in some regions. One further test specification was under development – TS 26.274, the conformance specification for the wideband version of the AMR codec, but it was subsequently decided to increase the scope of TS 26.131 and TS 26.132 to cover both codecs, and this specification was withdrawn.

There are two further related specifications: TS 26.074 and TS 26.174. These consist of a small document and an electronic distribution of a compressed archive containing a number of test vectors or binary sequences. These do not form part of the formal testing of the codec but can be used to verify that the implementation is bit exact. There is also ongoing work to add further codecs and associated test sequences as the multimedia capabilities of terminals and networks increase.

5.6 Main Organizations Involved and Their Aims

The 3GPP is the body responsible for UMTS standardization. The partnership referred to in its title is a partnership of six regional standards organizations:

- ARIB, from Japan
- TTA, also from Japan
- The T1P1 committee of ATIS, from the United States
- TTA, the Korean standards body
- CCSA, the Chinese standards body
- ETSI, the European representative, which has a special role. It is the hosting partner, providing the physical location for 3GPP in its headquarters in Sophia Antipolis in the South of France, the secretarial resources for the operation of 3GPP and the corporate infrastructure (IT, finance and so on).

The working groups themselves are mainly made up of representatives from the industry. Any company that is a member of one of the organizational partners is entitled to send representatives to the standards meetings. A working group has a chairperson, who will be from an attending company, and is voted into office by the regular attendees to the working group meetings. It will also have at least two vice-chairs, which is elected in a similar manner, and a secretary provided by ETSI.

The scope of 3GPP is the development of the standard, including the conformance tests required to prove that a terminal conforms to the standard. It is not responsible for providing type approval, or in any way regulating the introduction or use of terminals on real networks. Although it is made up from regional standards organizations, it is not a regulatory body in its own right. The approach taken is that the standards forum generates a library of conformance tests. Those bodies that do have responsibility for type approval or certification of terminals can then select from this library to create a test regime.

The first such body to do this was the GCF. This organization started life as a forum within the GSM Association. It was originally titled the GSM Certification Forum and was created in the wake of European deregulation to establish a self-certification regime. Early in the standardization process the GCF expanded its scope to cover UMTS certification and started selecting conformance tests from the growing library developing within 3GPP.

Under deregulation, type approval is mainly the responsibility of terminal manufacturers. The intention is that they self-certify their equipment in whichever way they feel most appropriate. However, in practice the GCF regime is widely accepted by manufacturers and most importantly by network operators. In most regions they are the principal customer of the terminal manufacturers and often require some assurance that a terminal has passed the GCF regime before accepting it. In Europe the network operators all recognize the GCF certification, and while in many cases they may require some additional level of testing specific to their network, there are no alternative or competing type approval regimes. GCF certification always forms a basic part of any operator's requirements. The advantage for the manufacturers is that by following a single regime and a single certification exercise, they can combine a significant part of the acceptance testing work load needed to deal with the European operators. This has given considerable weight to the GCF certification process, and has established the GCF as the main authority on terminal certification. Many other regions now look to the GCF as a reference, either indirectly, by following their selection of test cases, or in some cases directly, by simply accepting GCF certification as valid within their area.

The PCS Type Certification Review Board (PTCRB), which performs a similar role to that of the GCF for the North American region, uses the GCF as an indirect reference. This is largely because the frequency band allocations in North America are different from the main UMTS deployments in other regions. Where the test cases are independent of frequency band, and this is the large majority, the PTCRB follows the GCF selection but the tests are run within the North American frequency bands. With North American operators, certification by the PTCRB forms a similar basic requirement for acceptance of terminals.

5.7 Process

5.7.1 Specifications, Versions and Change Control

At the core of any open standardization process is a system of document control that, once a specification has reached some level of formal agreement, provides a mechanism for both agreeing changes to the document and traceability of the changes and the reasons behind them. The 3GPP process was imported from ETSI, where it has been in operation for a long time, and is very effective. This process is worth outlining, as it applies equally to conformance test specifications.

The process starts with the creation of a work item, as described in Section 5.1. A work item will usually require either the creation of one or more specifications, or some modifications or extensions of existing specifications. New specifications are drafted within the responsible working group, and often by an SWG or ad-hoc group created specifically for that purpose. Once the specification reaches a reasonable level of content – 3GPP guidelines recommend at least 50% of the planned contents – it is presented back to the working group for approval to version 1.0.0, and then by the working group to the TSG for information. The drafting work will continue on the specification, without formal change control, but now the evolving document, in its latest state, will be presented back to each working group meeting for approval and wider circulation. Each time this happens, the second digit of the version number is incremented by one. Once the specification reaches around 80% of the planned contents and those contents are reasonably stable, it is presented to the plenary meeting of the working group's parent (e.g. TSG-RAN or TSG-CT). If it is approved by the plenary, it is raised to version 2.0.0, and then presented at the TSG plenary meeting for approval. If the specification is approved, the first digit of the version number is set to the release of 3GPP, which this specification is targeted at, with '3' representing Release 99. New specifications being approved in 3GPP at the time of writing are generally transitioning from 2.0.0 to 8.0.0 once they have TSG approval. Following this transition, the specification enters formal change control. From this point, anyone who wants to make changes or add to the document has to submit a change request at a working group meeting. The change request contains a header form with an explanation of why the change is required and summary of what the change is, and it contains a copy of the area of the specification in which the change is to be made, marked up with the change itself. This is done using a widely available document processing application, Microsoft Word, which has the capability to track changes using revision marks. Each working group has an email reflector which allows members to distribute documents by submitting them to one email address, and through the reflector change requests are generally distributed shortly before a working group meeting. At the meeting itself the change request is presented and justified usually by the author, or by another interested party, and debated by the meeting attendees. Generally approval is by consensus. That is, no one actively opposes approving the change. Often consensus is reached through compromise, and change requests may go through many revisions in order to get acceptance by everyone. A working group meeting will often consider many change requests to a specification, and part of the work of the meeting is to resolve conflicts and overlaps between change requests, and by the end of the meeting there will be a list of approved change requests for each specification. The working group approved change requests are then grouped together into logical batches, and presented to the TSG plenary for their approval.

Conformance tests are changed or added through the same process. Once the conformance specification has reached formal change control, the addition of new tests, or the modification or correction of existing tests, requires a formal change request. This process is also applied, albeit in a slightly modified form, to changes to the TTCN test case implementations. As TTCN is essentially software, the process has needed a little adaptation.

5.7.2 TTCN Test Cases (Signalling)

3GPP took the approach of funding a team of testing specialists to write implementations of the protocol conformance tests. The resulting test suites are the reference for protocol testing and are freely available as part of the specification. ETSI were given the job of establishing this team and running it, and its official title is Task Force (TF)-160. The team is staffed by a full-time ETSI manager and then by experts from the industry who work in ETSI's Sophia Antipolis headquarters. The experts are either submitted on a funded basis, in that their employer receives some financial recompense for their services, or in some cases submitted on a voluntary basis, where their employer receives no recompense. The team size varies, depending on the work load and funding, but has usually around five or six people. This team writes TTCN test cases and performs all the formal updating of the existing ones. However, they have no way of debugging the tests they write. This task requires specialized test equipment, and close partnerships with the UE developers, as test cases are usually being debugged as the UE functionality is being developed. This may seem to be a strange way to develop test software, but actually works remarkably well. The team works very closely with the major test equipment manufacturers, who carry out the debugging, reporting the problems they find back to TF-160, who fix the official TTCN.

The process for approval of a test case works as follows. Once a test case has been drafted by TF-160, it is released to the RAN5 members for review. It is the test equipment manufacturers who are mainly interested in it as their business is selling the test systems needed to run the tests. They will take the test case and start to debug it. Once they have it working against a UE, they will submit a change request with the changes needed to get the test case running. This CR will contain marked up changes to the TTCN and usually needs both special tools to view it and specialist knowledge to understand it. Therefore it is usually distributed on the reflector when it is ready and is not handled in the working group meetings. The change request also has some additional information to allow other parties to review the change adequately. It is usually accompanied by the protocol logs taken from the test equipment when the test case was successfully run, together with information to show how the test system was configured. This package of information then goes through an email approval cycle where if no objections to the changes are made within a given time period, the change request is considered approved and is implemented by TF-160 in the next appropriate release of the test suite. The time allowed for email approval depends on two factors. It is normally 4 weeks, but this is reduced to 2 weeks if the test case has been shown to run successfully on two or more independent UE implementations, or if the test case is similar to one that has already been approved on the same test platform. Subsequent changes to the test case follow a similar cycle, but the review cycle is always 2 weeks.

5.7.3 The GCF and Test Case Validation

It has to be understood that the major goal of the process of approval for test cases in RAN5 is to bring the test case under change control. At this point the test case is still likely to have some errors in it. Test cases are designed to cope with any possible set of UE options and capabilities and therefore contain many branches. At the point of approval, many of the paths through the test case will not have been exercised. The certification programme run by the GCF is partly based on the conformance tests approved by RAN5, taking a fairly large selection of about 500 test cases covering those areas that the forum considers to be most important. The GCF divides logical groupings of test cases into work items (not to be confused with 3GPP work items), where typically each work item will cover a different feature or technology. For example, GSM Release 99 is Work Item 005 (WI-005), and

UMTS HSDPA (Release 5) is WI-014. To accelerate the introduction of UMTS, the original UMTS FDD-mode Release 99 tests were prioritized into two work items, with the high priority tests going into WI-010 and the lower priority into WI-012. A list of relevant UMTS work items is given in Table 5.4.

Table 5.4 Relationship between GCF work items and 3GPP conformance specifications

Work Item	Description	3GPP Specifications
WI-010	UMTS Release 99 FDD high-priority tests	34.121, 34.123
WI-012	UMTS Release 99 FDD enhancements (second priority tests)	34.121, 34.123
WI-013	UMTS Release 4 and Release 5 (non-HSDPA)	34.121, 34.123
WI-014	HSDPA (Release 5)	34.121, 34.123
WI-015	A-GPS (Release 99)	34.123
WI-024	UMTS Release 6 FDD enhancements (non-HSUPA)	34.121, 34.123
WI-025	HSUPA [enhanced uplink (E-UL)]	34.121, 34.123
WI-030	A-GPS Release 6 minimum requirements	34.171
WI-031	Integrated Multimedia Subsystem (IMS) call control (Releases 5 and 6)	34.229
WI-035	USAT conformance testing	31.124
WI-049	Multimedia Broadcast and Multicast Service (MBMS)	34.121, 34.123

Each work item relates to a list of test cases selected from the listed test specifications. UE manufacturers have to pass the tests relevant to their terminal within a work item, and they must do this for each work item relevant to their terminal. To use test cases for certification, the industry must have a high degree of confidence in them, so the GCF uses a rating system to indicate when test cases reach a suitable level of stability and maturity. Test cases go through a validation process and, depending on the level of validation, are assigned a category. Only certain categories can be used to certify terminals. There are also rules about when a work item is itself mature enough to be used for certification. The GCF maintains a database of tests and their validation status that manufacturers can use to determine what testing is required for certification. This is known as the Certification Criteria (CC) database.

For a test case to become accepted for use in certification it needs to be validated. In essence, this means that an independent and competent body has adjudged the test case to fully meet the purpose for which it was intended. Validation is usually performed by test houses, or businesses that specialize in this type of testing, and the GCF maintains a register of test houses that they recognize as competent. This is determined through a system of accreditation. The accredited test house is required to run the test case against at least two independent commercial UE implementations using a commercially available test platform and, through reviewing the test case and the execution logs, determine whether it is valid and which of the various validation categories it falls into. The test house will then submit a validation report to the GCF. If the report is approved then the test case is validated on the test platform used by the test house. The result of this is a matrix that for each test case will show which of the test platforms it is validated on.

5.7.4 The PCS Type Certification Review Board

The GCF has a counterpart organization in North America that in the past has taken care of GSM-based PCS 1900 certification, but has now extended its role to cover UMTS. The PTCRB works very closely with the GCF, has a similar process and reuses much of the testing regime set by the GCF. The main difference is that, as with GSM, the North American networks operate at different frequency bands to the European operations, so even if a UE has passed GCF certification it still needs to gain PTCRB certification to be accepted by North American operators. Like the GCF, the PTCRB has a validation process, which incidentally uses a very similar categorization, and requires tests to be validated by accredited test houses on commercially available test platforms. Again the PTCRB maintains a matrix that shows for each test case which test platforms it has been validated on.

5.8 Certification

Certification is the equivalent of type approval but in a deregulated environment, the main difference being a shift in emphasis from a legal imperative to a market imperative. Some regulation still exists though, and this varies from region to region. This legislation tends to be mostly focussed on protection of consumers and protection of other users of the spectrum. In Europe this comes from the EC.

5.8.1 The R&TTE Directive and CE Marking

Despite its evolution to a more global organization, the GCF is still predominantly driven by European legislation, and the processes it has developed for certification originate from EU directives, in this case mainly from the 1999 directive 99/5/EC, also more usually known as the Radio and Telecommunications Terminal Equipment (or R&TTE) directive. This directive replaces the various type approval processes that existed in different European states, and provides the basis for 'CE marking' of equipment in Europe. The R&TTE directive is principally concerned with health and safety, EMC and the fact that the radio spectrum is used within the terms of the appropriate license; however, there are also some additional requirements which may apply to UEs covering proper use of network resources, personal and data privacy, proof against fraud and access to emergency services. For radio telecommunications UE, the R&TTE directive gives three levels of diligence the manufacturer can apply to achieve compliance.

1. Officially called 'Internal Production Control plus specific apparatus tests', this is the most basic form of self-certification. In this case, the manufacturers do all the work of proving and documenting compliance themselves. They must provide a written 'Declaration of Conformity' and provide with it technical documentation with enough detail to allow an independent examiner to establish conformance. This means it has to contain considerable detail, such as design documents, manufacturing drawings, circuit diagrams,

and descriptions of the operation of the product, and most importantly from the perspective of testing, details of what tests have been carried out, including test reports. For the actual tests that need to be carried out, the directive states that the manufacturer must have applied the 'essential' tests. The determination of what tests are essential is delegated by the EC to what it terms 'notified bodies'. These are organizations throughout the EU that are appointed under strict conditions by a notifying authority for their country, which is usually a government department or ministry. The notified bodies are generally test houses, often the same ones that are accredited by the GCF.

2. The next level is incremental in that the manufacturer has to do everything explained in the previous paragraph, and in addition he had to produce something called a 'Technical Construction File (TCF)' demonstrating how he proved conformity. This TCF is largely a more formal version of the technical documentation above, but in this case the TCF is submitted to a notified body. They will review it and provide an opinion to the manufacturer as to whether it proves conformity; however in accord with the principles of self-certification, it is still the manufacturer who makes the final decision on conformity.

3. The top level is called 'Full Quality Assurance'. At this level the manufacturer has to operate a quality system that runs through design and development, testing and production. The quality system is reviewed by a notified body to ensure that all the manufacturer's products meet the requirements of the R&TTE directive, and then the system has to be regularly audited, and this can include unannounced visits. The quality system has to provide quite a high level of formal documentation showing what testing has been done both during development of the product and on an ongoing basis, during its manufacture.

Most importantly for all these options, this is an ongoing process; the manufacturer also has to ensure continued compliance, and has to demonstrate for any change in the product that it is still compliant. These three options correspond to Annexes III, IV and V of the directive. There is also an Annex II which has a lesser obligation and does not require the involvement of a notified body; however this cannot be used with a radio transmitter.

As we move up these levels, there is an increasing cost associated with the increasing involvement of an external party, the notified body. However, this cost has to be balanced against risk. Each EU state has an agency responsible for enforcing the directive, and they will usually have inspectors who check products once they go on sale in that country. The penalties can be quite severe if the product is found to be noncompliant, so the stronger the proof of compliance is the better. Furthermore, full quality assurance is a form of certification that operates on a systematic basis at a more fundamental level than an individual product. Provided that each new product follows the approved system, it does not individually need to be audited by a notified body. For larger manufacturers with a continual stream of products coming to market, this method is much more efficient, and although it may be more costly at the start, it be more economical in the long run.

5.8.2 The United States and Japan

In the United States it is the Federal Communications Commission (FCC) that has responsibility for authorizing cellular telephony devices. The particular piece of legislation is the Code of Federal Regulations (CFR), Title 47. The FCC regulations are quite broad, but two key parts relate to the amount of radiation absorbed by the human body, known as SAR and EMC. The SAR requirements come from the Institute of Electrical and Electronics Engineers (IEEE) standard C95.1–1991, 'IEEE Standards for Safety Levels with Respect to Human Exposure to Radio Frequency Electromagnetic Fields, 3 kHz to 300 GHz', and the EMC requirements come from the CISPR standard CISPR 22. The process here relies on using a system of certified laboratories to perform the testing and certification.

In Japan the regulation is set by the Radio Law, which covers all use of the radio spectrum, but specifically article 38 relates to the certification of terminals. The Ministry of Internal Affairs and Communications (MIC) is responsible for administering the law, and the act gives it the scope to appoint Designated Certification Agencies, who are authorized to inspect equipment. Much of the certification is performed by the official government certification agency, the Telecom Engineering Centre, or TELEC. The standards that are applied come from the ARIB, mainly from ARIB's T-57 standard, which itself heavily references CISPR 22.

5.8.3 GCF Certification

The regulatory testing lays down a basic framework to ensure that terminals do not harm people or disrupt other services or equipment. Beyond this is a need by the industry to make sure that users experience a satisfactory level of quality. There are a variety of drivers for this. In particular, when dealing with networks with millions of subscribers, technical problems with the operation of UEs become magnified, and logistically much more difficult to manage, and can be costly. There is a need for a higher level of certification than just the basics required by law, and this is widely recognized in the industry. The two main organizations offering this higher level of certification are the GCF and PTCRB (Section 5.3), and both have similar processes for achieving it. Whilst not mandated by law, these certifications tend to form an entry point to gain acceptance by operators to retail the handset, and therefore are as important as the statutory certifications from a business sense.

The GCF process starts with GCF membership; to submit a product for certification, the manufacturer has to be a 'Quality Qualified' member of the forum. Becoming quality certified requires the manufacturer to follow a recognized quality programme, such as the International Standards Organisation's ISO 9000. As is usual with such quality programmes, they have to be regularly assessed by an independent auditor. To submit a device for certification, it has to pass a large set of tests known as the CC. The testing has to be done in an accredited laboratory, where the most widely recognized form of accreditation is ISO 17025. The CC actually go beyond the official 3GPP conformance tests (Table 5.4), covering the following areas:

- 3GPP protocol and RF conformance tests
- 3GPP USIM tests
- 3GPP audio codec tests
- 3GPP A-GPS conformance and performance tests

- Application enablers tests, including the following:
  - Multimedia Messaging Service (MMS)
  - Push-to-talk over Cellular (PoC)
  - Internet Messaging and Presence Service (IMPS) [essentially Instant Messaging (IM)]
  - Videotelephony
  - Web browsing
  - Secure User Plane for Location (SUPL).
- Field testing.

For the GCF process, the accredited test laboratories do not need to be independent of the manufacturers. They may declare that they have the necessary means to carry out the testing themselves, but accreditation is still required, and a number of larger manufacturers have their own accredited test laboratories. Once the UE has passed the relevant CC tests, the results are compiled into a compliance folder and this folder forms the supporting evidence for the manufacturer's declaration of certification. As in legislative testing, it is the manufacturer who ultimately has responsibility to declare compliance. One important point is that the CC database contains many validated tests, each having been validated on at least one test platform. However, for the test itself to be considered valid, it must be run on a test platform for which the test case has been validated. This means that particularly for the newer work items the test laboratory may need to use more than one platform to cover all the test cases relevant to the UE. The GCF process does have a little flexibility and permits four alternative approaches to certification:

- The CC database actually forms a reference. The manufacturer can choose to follow exactly the tests listed in the database, and this is the most widely used approach. However the manufacturer can:
- use an alternative test regime provided that he can demonstrate that it results in the same level of confidence as the CC database, or
- not follow a specific test regime, but be able to demonstrate that whatever testing has been done yields the same level of confidence as the CC database, or
- not carry out any specific testing, but be able to demonstrate by analysis of the design and manufacture that the UE is compliant, again to the same level of confidence as the CC database.

The catch in all of these is proving that the alternative process gives the same level of confidence as the CC database. In practice, the middle two alternatives are mainly there to allow other formal test regimes to be recognized, and their importance is mainly in allowing more effective sharing of effort for manufacturers who want to get both GCF and PTCRB certification. The last option is mainly provided for UEs that are rebadged or branded from an already certified design. However, this needs to be treated with some care. UEs are extremely complex systems, and small, and sometimes innocuous looking changes in the software can have much broader effects. Changes to the user interface or applications on the phone need to be reviewed carefully to make sure they have not impacted complex relationships and timings between software modules that could effect compliance. In general, most changes to the original UE design will require some level of retesting.

5.8.4 The PTCRB

The wideband CDMA access network was originally developed to meet the IMT-2000 initiative, which was so named because the spectrum allocated sat around 2000 MHz, actually either side, with the uplink operating between 1920 and 1980 MHz and the downlink between 2110 and 2170 MHz. Because of the global proliferation of networks based on the standard and the vision of global roaming of handsets, as regions look to auction off new spectrum, the standard has been steadily expanded to cover this. The core technology itself is quite independent of the carrier frequency, so applying it to new frequency bands is mainly a process of reworking the RF characteristics. The United States and Canada use band II (1850–1910 MHz uplink and 1930–1990 MHz downlink), nominally called 1900 MHz and band V (824–849 MHz uplink and 869–894 MHz downlink). More recently, the US government has allocated spectrum at 1710–1755 MHz uplink and 2110–2155 MHz downlink, known as band IV. The North American operators therefore cannot directly reuse GCF certification, so the PTCRB has created a test regime in these frequency bands. The actual test cases are basically made from the same selection of 3GPP conformance tests made by the GCF, the difference being that these tests are run within the North American bands applicable to the terminal. This means that if it is multiband the testing needs to be applied in each relevant band. The PTCRB also maintains a list for each test case of the test platforms on which it has been validated and for which frequency band, and the test laboratory has to run the test on a platform that has been validated for that band. As the test cases are materially the same, and a significant part of the testing time is in setting up the correct environment for testing, having test systems that can easily switch between frequency bands can provide a real gain in efficiency where terminals need both GCF and PTCRB certification. There is one further important difference. The GCF certification regime requires manufacturers to perform some field testing, usually in a number of live networks, but field testing is not part of the PTCRB regime.

6

Interoperability Testing

6.1 What is Interoperability Testing?

On the surface, interoperability testing and conformance testing are both trying to do the same thing. In theory, if we have two implementations that conform exactly to the same standard interfaces, then they should interoperate correctly. So, it is worth looking at why manufacturers spend so much time and effort testing against real-infrastructure equipment. Is this a failing of conformance testing? The answer is that, in practice, it is not realistic to create a set of conformance tests that provide a high enough level of test coverage to provide a guarantee of interworking. This is particularly the case for a complex standard like UMTS, where there are so many different paths for implementers to take. This does not mean that conformance testing does not have a role to play, but it does mean that something else is needed to complement the conformance tests. Conformance testing performs three critical and complimentary roles:

1. It provides a reference baseline 'behaviour' that infrastructure and network designers can rely on having been tested.
2. It tests behaviours and procedures which are not yet implemented in network equipment.
3. It tests the UE in the context of the system. Interoperability is very much focussed just on the functionality of the UE being tested. However, especially with WCDMA technology, devices that behave 'badly' can appear to function well in themselves, but can destroy capacity on the network, or interfere with or reduce the level of service available for other users in the cell, and this can be very difficult to detect from interoperability testing.

In doing this, conformance testing is essentially trying to future proof the network. Interoperability testing, however, is focussed on the 'now'. It tests against infrastructure as it currently is, including current software versions, current behaviours and any quirks of operation. Its strength is that it gives the truest guide that the UE will operate with the tested

Testing UMTS: Assuring Conformance and Quality of UMTS User Equipment   Dan Fox
© 2008 John Wiley & Sons, Ltd

configurations. Its weakness is that any changes to the infrastructure equipment, including software updates, can still bring hidden UE problems to the surface. This is especially the case in early network deployments where the infrastructure software is rapidly evolving and there are frequent updates.

Interoperability testing often starts very early in the development cycle and through a number of routes. Some manufacturers have close partnerships with or even sister organizations that are infrastructure developers. Once the equipment on both sides is ready to support a feature, then it can be beneficial both ways to start testing together. Although simulators are often available with similar support, testing with infrastructure provides another independent verification and can often bring to light a different set of problems. Another method is to use a test network operating in the development facility. However for features which are very new, this still requires a close relationship with the infrastructure developer as the features may not be available in commercial software builds. Both these tend to be the preserve of larger manufacturers. In particular, setting up and running a test network can be very costly. For other manufacturers, interoperability testing presents a much greater challenge. This is particularly the case for companies supplying components and subsystems, and there is increasing competitive pressure for chipsets and protocol stacks to have already had some level of interoperability testing.

6.1.1 Interoperability Laboratories

There are a number of interoperability laboratories running test networks with real-infrastructure equipment. They tend to be run either by infrastructure vendors or by network operators. There are several difficulties with using these laboratories, and as a result they have to be used with careful planning. Some of the laboratories are operated commercially and can be accessed at any time during the development process. However, there is usually a significant cost associated with using the laboratory. Others, and this often applies to operator laboratories, are run for specific purposes, such as acceptance testing of UEs for resale by the operator. In this case, it can be very difficult to get access to the laboratory until very late in the development cycle, when the UE is close to commercial deployment. Regardless of the type of interoperability laboratory, there are some general issues with using them:

- The demand for interoperability testing has grown considerably over the past years, and the laboratories are often very busy with long waiting times to get a slot.
- When the slot arrives, the test team usually has to follow a prepared test agenda, as some of the testing will require the network to be configured or operated in a particular way. This requires coordination with the operating staff, and therefore the test plan has to be prepared and agreed in advance and then followed.
- Unless the test team are fortunate enough to be sited close to a suitable laboratory, access often presents the logistic challenge of getting equipment and engineers to the laboratory for the duration of the test slot.
- Because time is expensive and limited, it is not practical to debug problems on site. The normal approach is to run as many tests as possible, regardless of outcome, and then take the results and test logs back to base for later analysis.

These issues, particularly the last one, tend to limit the use of this type of interoperability testing early in the development cycle for smaller manufacturers. If the UE software is not already fairly well debugged, then the chances are that a few systematic problems will prevent large numbers of test cases running, masking many other issues and limiting the usefulness of the session.

6.1.2 Simulators and Interoperability

One of the most successful and effective ways of carrying out interoperability testing is to combine the use of system simulators with infrastructure testing. Many of the laboratories will provide detailed test specifications showing the signalling flows for each test, together with key parameter settings. These test specifications can be converted into simulator test cases that follow closely the laboratory tests. In practice, the simulator cannot replicate the exact behaviour of the infrastructure. The effort required to match all the timings and every parameter would be very large, and in any case, this level of information is very difficult to get. However, the simulator tests can get close enough to give a good level of confidence. This can be very effective at eliminating systematic problems before the interoperability session starts and leads to a session which is much more productive. A simulator can also prove useful after a test session. The test logs and other data may help to determine what went wrong when a test failed, but it does not necessarily show why things went wrong. Recreating failed tests back in the developers own laboratories can be a very powerful way of rapidly debugging problems. The basic scenario may already be available from the preparatory testing done on the simulator. With the additional information from the logs, the simulator test can be progressively refined until the point where the failure is replicated, at which point the cause of the problem is usually much easier to identify and fix.

6.2 Interoperability and Certification

As a development cycle moves closer to the commercial release of the product, there is increasing pressure to make sure that the UE works with current live networks. If there are problems when it initially gets out into the market, the operators may be swamped with customer service calls and the manufacturer inundated with product returns. Even if the UE only fails once in 1000 calls, a million subscribers each making 10 calls will generate 10 000 failures. Interoperability testing can act as an important gate to the commercial deployment. Problems found before the UE starts to be used can be easily dealt with, but problems that emerge once the UE is in widespread use are very much more difficult to deal with. To some degree, if they emerge when infrastructure software is updated, or new services added, they can sometimes be mitigated by delaying the roll-out of the update or service or by working around the problem from the network side. However, even though some UEs nowadays can be reloaded with updated software, logistically this is very difficult to do once they are out on the streets, and if a network side solution cannot be found, they may effectively be unfixable. Hence, both interoperability and conformance testing play a significant role in certification and operator acceptance testing.

6.3 Ways in Which Interoperability Testing is Carried Out

6.3.1 Live Network Testing

The GCF certification criteria (Section 5.5) include an amount of interoperability testing that is expected to be performed on live networks. This is covered in the field trials work items (listed in Table 6.1). A field trial has advantages and disadvantages over a test network. As testing is performed on a real network, it provides the ultimate level of confidence that at least the basic procedures in the UE will work correctly with that network configuration, but balanced against this are a number of limitations. Live networks cannot be reconfigured or otherwise coerced to produce negative or failure scenarios. Neither can they be set up to create boundary, or worst case, conditions for the UE (although sometimes there are geographic 'hot-spots' which can create useful test environments), and these are often where problems that have slipped through development and conformance testing lie. Testing on a live network is also difficult to control in a repeatable way. The environmental conditions (e.g. how many other nearby users there were, strengths of neighbour cells and so on) vary outside control of the testers, and test results can be very difficult to replicate. Furthermore, it can be very difficult to determine the true network conditions at the time of the test, and it can be difficult to apportion blame for some types of failure. For example, if a call failed due to the network being busy, was the network genuinely busy or was this a symptom of a UE problem? Nevertheless, live network testing is relatively easy to carry out and does have value, so it does form an important part of interoperability testing.

Table 6.1   GCF current field trial-based CC

WI-027	Field trials for HSDPA (FDD)
WI-029	Field trials for Release 99 UMTS (FDD)

Some examples of the types of tests that can be done on live networks are:

- mobile-originated and mobile-terminated speech calls
- calls to a busy number
- supplementary services (call forwarding, call waiting, etc.)
- multiparty calls
- video calls
- packet-switched data sessions
- handovers and reselections
- SMS
- MMS
- specific UE functions, such as PIN handling and unblocking.

In these cases, typically only success cases can be tested. The live network can also be used to perform throughput testing for packet-switched data, which particularly for 3G services is one of the main selling points. However, the results need to be interpreted with care. There are many other potential bottlenecks within the network, and low throughput may not necessarily be a UE issue.

6.3.2 Test Networks

A test network can overcome some of the disadvantages of a live network, in that more aggressive testing can be done. Some of the environments that are difficult to find in a live network, such as barred cells, out-of-service cells, roaming environments with multiple network country codes, networks with different feature support and so on, can be created with a test network. They still have the advantage of being very close to the real environment, and this is especially true for some of the test networks run by network operators, which can accurately represent parts of their real network. However, they still have a couple of disadvantages or limitations. They are expensive to set up and run as they require a reasonable amount of network infrastructure, much of which is dimensioned for handling large numbers of users, even when scaled to a minimum. They are also often placed in screened rooms to prevent interference with real networks. One key limitation is that whilst they do offer more flexibility than a live network, there are still many scenarios that are difficult to create on a test network, either because of the physical environment or because there are inherent difficulties in dynamically reconfiguring the network elements during a test. Another problem for test networks is that it is difficult to get repeatability of test results, and in cases where the results are doubtful or indeterminate, it can be hard to establish whether a problem really existed. This can make it hard to catch intermittent problems, for example. As with live networks, in the end the benefits are significant, and a number of operators and network equipment vendors run test networks, and these are often used both for UE manufacturer's development testing and for operator's UE acceptance testing.

6.3.3 System Simulators

We have already seen that dedicated test systems, simulators, can be used effectively to aid interoperability testing during development. However, there is an increasing trend to use simulators directly as part of interoperability test programmes, particularly for operator UE acceptance testing. Simulators have the advantage that they are very flexible, and the environment they create for the UE is very controllable. They can be programmed to look like any network environment, with full control over each cell configuration, including physical parameters such as its power level. But more importantly, they can be reconfigured very rapidly, even during a test case. This opens the door to a wide variety of potentially important test scenarios. It is also easy to generate all the possible failure scenarios, including multiple level failures, such as failures during recovery procedures. Another big advantage of simulators is that tests tend to be scripted and are therefore highly repeatable. UEs are complex radio systems, and occasionally tests will fail due to external factors even though there is no underlying problem in the UE. A simulator can usually separate out these types of spurious failures from genuine intermittent faults. However, simulators also have their own set of disadvantages. They are highly specialized systems and require a level of expertise to create the test scripts. Each test needs to be carefully planned, designed, written and verified. This is a resource-intensive task and needs specialist engineers. Once tests are written, then of course they can be used indefinitely, but still a small amount of effort is needed to maintain them as the standards evolve. The other major problem is that while simulators are highly programmable, in practice it is very difficult to make them look and behave exactly like a specific network. There are two reasons for this. One is that, as in any test system, they are a compromise between cost and functionality. One of the main balances for a simulator is to provide enough test capability whilst limiting the amount of hardware. This means in practice that they have technical limitations which are not present in networks. An example of this is that a network may be configured with several common control channels (S-CCCH). A simulator will probably have enough hardware to support this configuration on one or two cells, but if the test requires many cells, the simulator will eventually run out of hardware. The second reason comes from the complexity of UMTS. There are so many parameters that characterize the signalling and configuration of the network that to replicate them exactly in a test case is a huge task. Once more, the disadvantages have to be seen in context. The flexibility of simulators makes them a compelling tool, and more and more operators and UE manufacturers are now using simulator test suites as part of their acceptance and development programmes. In fact, one stimulus for the growing use of simulators is that a lot of work has been done to overcome some of the disadvantages. Some of the simulators now have much easier to use scripting environments, including 'drag-and-drop' graphical development tools that require far less specialist knowledge to use. To complement the increasing ease of use, some of the specialist telecommunications service companies now offer cost-effective commercial services to develop custom interoperability test suites on the more widely used simulators.

6.3.4 Simulators and Network Emulators

A network emulator in concept is a box that behaves exactly like a real network. That is, unlike a simulator, it does not need a script to programme its response to UE signalling. The emulator is programmed to respond to any signalling received from the UE in exactly the same way a network would. It also has the capability to initiate events to the UE interactively under user control, for example, through various controls on a graphical user interface. Emulators are actually simulators with an extensive script, usually in the form of an executable programme, tied to a graphical user interface to provide user control. Of course, pure emulation leads straight back to the limitations of a test network, so they provide some capability to modify the behaviour of the emulation, both through scripts that operate at a much higher level than a simulator script and through extended control from the user interface. Emulators fill the gap between test networks and simulators. They are not as accurate to real network operation as a test network but provide more flexibility, and they are not as flexible as simulators but are easier to operate and do not require scripting. The main disadvantage for network emulators lies in how well they can actually reproduce the network behaviour. UMTS network behaviour is very complex, and writing an emulator that can really follow all the possible variations of UE behaviour and all the abnormal situations that can commonly arise is a very large task. The task is magnified by the fact that interoperability testing is fundamentally about making sure the UE will interoperate with network equipment from different manufacturers. Therefore, the network emulator actually needs to emulate a range of different network configurations to be really useful, and the behaviours can be significantly different. This is compounded by the continual evolution of the standards, requiring that new features are continually added to the network emulator. The large amount of work needed to add new features to an already complex emulation may mean that the feature is not available when it is really needed, and this can further limit an emulator's usefulness.

6.4 Typical Sources of Tests

In the past, an interoperability test was often written as a set of instructions containing a sequence of actions for the test operator to carry out and the expected results from these actions. Tests are carried out manually, and the operator will check one of the human interfaces, such as the display or earpiece for the result. The trend towards using simulators or emulators for interoperability testing has meant that tests are more readily available in a form that can be executed on commercial test systems. Generally, interoperability tests are written using proprietary APIs or languages on the test equipment rather than using TTCN 2 and the conformance API. This is driven by a number of factors, but mostly, it is because interoperability suites do not need the formality of TTCN 2 and can therefore be developed more efficiently using other methodologies.

Interoperability tests come mainly from three sources:

1. Infrastructure vendors: Those who write test descriptions to help UE developers prepare for interoperability sessions on their infrastructure. These tests are not implemented on any commercial test systems and are generally not available commercially.
2. Network operators internal testing: There are a growing number of operators who develop or commission test cases that they use for acceptance testing of UEs. These tests are not commercially available but are used within the operator's own laboratories on commercial test equipment.
3. Test equipment vendor–supplied acceptance tests: A small number of network operators have established partnerships with test equipment vendors, and some fairly substantial libraries of acceptance tests are now commercially available. These tests can be used for general interoperability testing but also form the acceptance tests that the operator runs in his laboratories.

In addition to these, some UE manufacturers now develop their own internal libraries of interoperability tests which can then be used for regression testing and on future UE models.

7

Testing Beyond Development

7.1 Manufacturing Testing

Most of the complexity in testing a UE is in verifying the design. Once this verification has achieved a satisfactory level of confidence that the product works, it can be mass produced. This then brings us on to the testing that is done during production for verifying the manufacture of the device. Due to the high complexity of a modern UE and as the radio part invariably needs some set-up and adjustment to deal with manufacturing and component tolerances, each device needs a significant amount of testing during production. Except for a few niche areas, UEs are consumer products, manufactured in high volumes, and the assembly and testing costs form a key part of the overall product cost. Test time is the main component of the testing cost, and hence, there is always pressure to reduce this. This means that production tests tend to be a small subset of the design verification tests – those that can be carried out rapidly and will give the maximum confidence in verifying the manufacture.

The manufacturing process for a UE differs from company to company and also depends on the type of device: phone, PDA, PC data card and so on. Figure 7.1 shows an example of a flow that might be used in a mobile phone production line. From the radio perspective, the key parts of the process where testing is needed are shown in the oval. Typically, this will use an instrument known as a one-box tester. The name dates back to early phone manufacture, when production lines used a collection of instruments to perform the tests, but mainly centred on a signal analyser and a signal generator. A one-box tester combines these instruments into a single unit, together with a level of capability to perform network signalling with the UE.

Manufacturing of mobile phones is now a major industry and has created a significant demand for these one-box testers. As a result, they have become highly optimized towards their intended task, including fast measurement hardware to keep test times down. The testing itself is typically done fairly late in the production process, after mechanical assembly. This

Testing UMTS: Assuring Conformance and Quality of UMTS User Equipment   Dan Fox
© 2008 John Wiley & Sons, Ltd

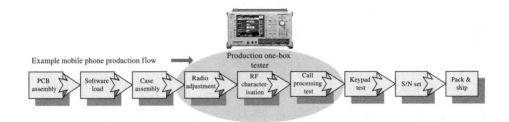

Figure 7.1 Use of testing in a UE production line (reproduced by permission of Anritsu)

presents a challenge of how to get the radio to transmit so that measurements can be made. At this stage, it is not possible to probe inside the UE, so some of the techniques that can be used in the laboratory, such as injecting signals or probing at strategic points in the transmitter or receiver chain, are not practical. The usual solution for this is to put a minimal network emulator inside the tester and use this to signal the UE into a call. Using the test loops already present for RF conformance testing, both transmitter and receiver measurements can be made. Signal analysers usually work by storing digitized samples in a memory buffer and then processing them. This allows many different measurements to be made from the same piece of captured signal. Measurements also need much greater precision than the typical demodulation processing of a receiver, which would need a large amount of signal processing hardware to do in real time. One-box testers therefore need two receiver paths: one which stores and processes samples in nonreal time, but with high precision, and one which can demodulate in real time, but with much lower precision. The following paragraphs discuss briefly the types of measurements made by the one-box tester and their relevance to the production process.

7.1.1 Radio Testing

UE radio transmitters need to use high-efficiency amplifiers, as pumping out RF power is one of the most costly things the UE does in terms of its total power budget. Unfortunately, the penalty is that these types of amplifiers are nonlinear. To work around this, various compensation measures can be applied to the signal before the amplifier stage to make it more linear at the amplifier output. However, this nonlinearity is sensitive to component and manufacturing tolerances, so to work out the correct compensation factors requires careful calibration of the transmitter. Following calibration, a set of tests generally based on the 3GPP RF conformance tests are then carried out on the UE. The full set of RF conformance tests can take days to run through, so a small subset is used to check the key characteristics, especially those that are most dependent on the calibration process. The main tests usually carried out in production are

- Frequency stability and carrier accuracy
- Maximum output power level
- Open loop power control of uplink transmissions
- Inner loop power control of the dedicated uplink physical channel
- Minimum output power level and power output when the transmitter is nominally off

- Occupied bandwidth
- Spectrum emission mask
- Adjacent channel leakage power ratio (ACLR)
- Transmitter modulation accuracy
- Peak code domain error
- Receiver reference sensitivity
- Maximum receiver input level
- Demodulation performance of the receiver.

These tests are essentially a subset of the RF conformance tests described in Chapter 8.

7.1.2 Call Processing Testing

Generally, the protocol stack is not extensively tested during manufacture, as it does not need any manufacturing adjustment. However, testing often involves some general confidence test using the call processing to exercise a reasonably substantial part of the UE. In some cases, this is combined with the radio measurements, in that call processing is used to activate the physical channels on which the measurements are performed, and interfrequency handovers can be used to repeat tests at different frequencies.

In recent years, there has been a tendency to try to reduce the amount of call processing used in the radio testing, as the signalling associated with setting up connections can take a significant amount of time in relationship with the overall test time. This can involve the addition of special test modes, usually proprietary in nature, that allow the radio to be activated without the need for all the associated signalling. However, in parallel with this there is also a trend to try to reduce signalling times in the networks, which may in the future counteract some of the benefits of testing the radio without call processing.

7.2 Service Testing

Service testing refers to the process of testing that takes place when a mobile device is suspected of being faulty once it is in service (i.e. in the hands of the end-user). This normally starts with the user taking the terminal back to a retail outlet or specialist service centre. From this point, a chain of progressive filters, each based on more extensive testing, is applied to weed out problems with usage of the device from failures relating to faults with its hardware. At the first level, a simple go/no go test is applied to the UE. At the very simplest level, this can simply involve attempting to make a call on the network; however, many faults are not as simple as this to identify, and a number of specialized go/no go testers are available (Figure 7.2). These will run a series of tests on the UE, including the setting up of basic connections, where a number of measurements are made of the characteristics of its radio receiver and transmitter. This type of tester will usually print out a short report which can be sent on with the device. Depending on the outcome of the go/no go test, the UE can be sent on for further repair or investigation, returned to the user as working correctly or returned into the supply chain for resale.

Figure 7.2   Commercial go/no go tester (Anritsu MT8510B, reproduced by permission of Anritsu)

7.2.1 UE Repair

Repair centres will often be able to make repairs to the mechanical parts of the UE, including the keypad, display, battery, internal cables and connectors, speaker and vibrator. Depending on the level of disassembly required during the repair, a simple go/no go test at the end of the process can provide enough confidence in the continued functioning of the device to allow it to be returned for use. More detailed repairs, such as replacement of circuit boards or of the antenna, can require a repeat of the full manufacturing line tests.

Part II

Testing by Layer

8

Testing the Physical Layer

8.1 Overview of the UMTS WCDMA Physical Layer

8.1.1 WCDMA and Power Control

The information capacity of a communication channel is dependent on three factors, related by the Hartley-Shannon equation, which defines the theoretical maximum capacity C of a channel in the presence of white noise as

$$C = W \log_2 \left(1 + \frac{S}{N}\right),$$

where W is the bandwidth of the information signal, S is the average power of the transmitted signal and N is the average power of the noise in the channel. WCDMA and other wideband, or spread spectrum techniques use a channel bandwidth W which is considerably greater than the desired information capacity C in order to support communication at lower signal to noise ratios. However, increasing the bandwidth also increases the average noise energy. WCDMA uses a technique called spreading to counter this effect and allow the noise to effectively appear only over the lower information bandwidth (see 'spreading' below). The technology originated in military applications where there was a need to transmit signals below the background noise level to avoid detection, or to receive signals in a high noise environment to provide immunity from jamming. For mobile telephony the same attributes are attractive, but for different reasons. The immunity to jamming is exploited to provide the ability for the channel capacity to be shared out over many users. To do this, the transmitted signal from each user is randomised and the energy spread evenly over the bandwidth so that it appears as noise to the other users. From the equation, while this is a very simplified view, it can be seen that as the noise N increases, for a given signal power S, the channel capacity available to the user will decrease inverse logarithmically. Put another way, the effect of adding more users into the cell is to reduce the available capacity per user, allowing

Testing UMTS: Assuring Conformance and Quality of UMTS User Equipment Dan Fox
© 2008 John Wiley & Sons, Ltd

cells to dynamically vary their operating point anywhere from having a few high data rate users to a lot of low data rate users.

8.1.2 Scrambling and Spreading

If each user's signal is truly randomized, then it would be impossible to recover it at the receiver. Fortunately, however, almost the same effect can be achieved using pseudorandomization. That is, an algorithm that generates a sequence of numbers that are distributed evenly throughout the number space but with no discernable pattern to their distribution. Because the number sequences are algorithmically generated, they can be independently but identically generated at the transmitter and receiver. These number sequences are known as pseudorandom number or PN sequences. This PN sequence can then be used to scramble the data stream before transmission. From a Boolean perspective, this involves performing an exclusive-or (XOR) of the data and the PN sequence bit by bit (Figure 8.1). The resulting data stream has the pseudorandom properties of the PN sequence but also contains the data. By repeating the XOR operation at the receiver, the original data are recovered.

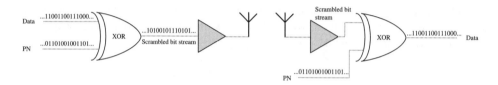

Figure 8.1 Boolean operations for scrambling and descrambling with pseudorandom number sequences

In the 3GPP specifications (TS 25.213), this operation is shown as a multiply. This is because of the way the data is coded, as we shall see shortly, which means that at this point the data stream is a series of +1s and −1s rather than 1s and 0s. Multiplying +1s and −1s has the same effect as a binary XOR.

Scrambling spreads the power in the transmitted signal evenly across the channel bandwidth, and ensures that each user's signal is completely uncorrelated with that of other users, so that it appears as white noise. In practical systems, the minimum acceptable signal-to-noise ratio (SNR) depends on many factors, but it cannot be allowed to become arbitrarily small, and with nearby users transmitting at similar power levels, the SNR soon reduces to a level where receivers cannot recover the data efficiently. The answer to this problem is provided by spreading. Suppose a communication channel has a maximum capacity of 3.84 Mbps, with a target data rate for each user of 240 kbps. For a single user, there is enough capacity to transmit 16 bits for each data bit to be sent. Doing this spreads the signal across the 3.84 Mbps channel from the original 240 kbps. In the receiver, it can be de-spread back to its original bandwidth by reversing the spreading process. This may seem quiet wasteful, as it now requires 3.84 Mbps to transmit a 240-kbps signal, but the advantage of doing this is that the effect of de-spreading is to concentrate the energy in the data signal over the smaller (original) bandwidth, but as the noise is not de-spread, its energy remains evenly spread over the original bandwidth. By filtering, the SNR can therefore be improved (Figure 8.2). This is known as coding gain, and the amount of gain is related to the ratio between the original bandwidth and the spread bandwidth. This is called the spreading factor (SF).

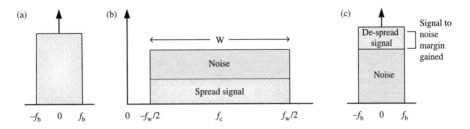

Figure 8.2 Spreading of the signal bandwidth to provide improved signal to noise ratio. (a) Original signal before spreading; (b) signal is spread, modulated onto the carrier and received with noise from other users; (c) signal is de-spread, demodulated and filtered. It now has an improved SNR

In its simplest form, we could spread simply by repeating each data bit (16 times in our example above). However, there is a more efficient way to spread the data using orthogonal codes. These are codes that have the property that any two codes in the family when multiplied together bitwise and the results of these multiplications are summed, the result is always zero. Put another way, codes in the family correlate completely with themselves and have zero crosscorrelation with any of the other codes. This can be seen if we take the following set of 8-bit orthogonal codes:

Code	Sequence							
0	−1	−1	−1	−1	−1	−1	−1	−1
1	−1	−1	−1	−1	+1	+1	+1	+1
2	−1	−1	+1	+1	+1	+1	−1	−1
3	−1	−1	+1	+1	−1	−1	+1	+1
4	−1	+1	+1	−1	−1	+1	+1	−1
5	−1	+1	+1	−1	+1	−1	−1	+1
6	−1	+1	−1	+1	+1	−1	+1	−1
7	−1	+1	−1	+1	−1	+1	−1	+1

If we multiply code 2 by code 5, for example, we get

$C_2 \times C_7$	
-1×-1	+1
$-1 \times +1$	−1
$+1 \times +1$	+1
$+1 \times -1$	−1
$+1 \times +1$	+1
$+1 \times -1$	−1
-1×-1	+1
$-1 \times +1$	−1
Total	0

This property also holds if we invert either or both of the codes:

$\bar{C}_2 \times C_7$	
$+1 \times -1$	-1
$+1 \times +1$	$+1$
$-1 \times +1$	-1
-1×-1	$+1$
$-1 \times +1$	-1
-1×-1	$+1$
$+1 \times -1$	-1
$+1 \times +1$	$+1$
Total	0

If we take two data bits in parallel, select code 2 for the first bit and code 7 for the second. If the data bit is a one, we invert the code, otherwise we take it directly. We have now spread our signal eightfold. However, we can add the two codes together in the transmitter, and then if we multiply the result by code 2, we will extract the first data bit, and if we multiply by code 7, we will extract the second data bit. This property can be used to create independent data channels within the spread signal. The 3GPP standard actually uses a tree of codes as shown in Figure 8.3, with the additional property that a code in the tree is orthogonal to all other codes except for those that are connected to it and to the right. That is, code $C_{4,0}$ is orthogonal to $C_{4,1}$, but it is also orthogonal to codes $C_{8,3}$, $C_{8,4}$ and to $C_{16,4}$, $C_{16,5}$ and so on. This provides a lot more flexibility in creating channels as mixtures containing both fast and slow channels that can easily be created.

Channelization codes are used slightly differently in the uplink and the downlink. On the downlink, the PN scrambling codes are used to distinguish each basestation, and the channelization codes are used to distinguish signals for different channels and users. On the uplink, the PN codes are used to separate out users, and the channelization codes are used to multiplex the different channels from one user. The main reason for this is that the zero crosscorrelation property of orthogonal codes only works if the codes are time aligned. If you redo the calculations above for C_2 and C_7, but offset them by one bit they start to correlate quite strongly. It is difficult to time align signals arriving at the basestation from different UEs, and the PN sequences show lower crosscorrelation when they are not closely time aligned.

8.1.3 Channel Coding

If we imagine a system with two signal states that represent a binary 1 and 0, once the noise level exceeds half the difference between the two signal states, we can no longer distinguish between a genuine 1 and a 0 with noise. Noise is random in nature, and so not every bit is going to be corrupted. Using forward error correction techniques, we can correct bits that are received incorrectly to some degree. The stronger the error correction, the lower is the SNR we can tolerate while still decoding the data. However, stronger error correction requires more bits dedicated to the error correction, and this reduces the channel rate for useful data. The 3GPP standard uses two coding schemes: a convolutional coder where half the bits are

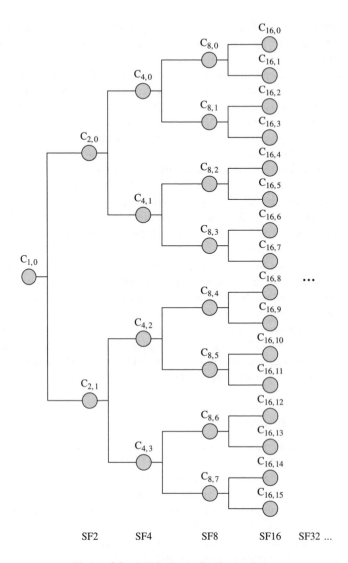

Figure 8.3   3GPP channelization code tree

useful data and half are for correction, and a turbo coder, where one third of the bits are for useful data and two thirds for correction. In addition, a puncturing scheme is used to provide more fine-grain control of the level of protection applied.

One further enhancement of the correction capabilities is provided by interleaving. In the simple model above, we assumed that the noise was random, and error correction techniques assume an equal probability of losing any bit during transmission. However, in real life, users move around and noise from other users and destructive interference from stray reflections of the user's own signal cause the noise level to vary considerably over periods of time much longer than a single bit period. This is known as a 'fade'. Interleaving spreads the bits out

evenly over a longer period of time, so that fades occurring within the interleaving period are made to look more like random noise. The network can choose interleaving periods of 10, 20, 40 or 80 ms.

If we extend the system beyond two signal states we reduce the distance between the states, and hence, we need a better SNR. However, there are advantages to doing this. For example, if we use a four-state scheme, we can transmit twice as much information in the same bandwidth. We need a better SNR to be able to do this, but particularly in a radio system, bandwidth is a precious commodity and to a point it is worth using modulation with higher orders of states. The 3GPP standard has four-state modulation for the Release 99 channels, a modulation scheme known as quadrature phase shift keying (QPSK). For the high-speed downlink channels (Release 5 and onwards), an adaptive scheme is used. If the actual SNR measured on the channel is high enough the transmitter can select higher orders of modulation: 16-state quadrature amplitude modulation (16 QAM) for Releases 5 and 6 and 64 QAM for Release 7. Release 7 also adds adaptive modulation using 16 QAM in the uplink for the high-speed channel.

The margin gained from the processing gain can be used in a number of ways. It can provide more capacity on the network (more users). This increases the noise, pushing up the bottom of the margin window. It can provide more original signal bandwidth, reducing the SF and pulling down the top of the margin window. Alternatively, it can be used to provide more quantization and therefore increasing the data rate within the same bandwidth. Here, we have an equation trying to balance three variables. Reducing the SF and increasing the number of users shrink the margin window down, but the channel coding and modulation sets the limit to how small the window can be compressed.

8.1.4 Channels and Frames

The use of channelization codes allows the radio channel to be divided into code channels, either serving different purposes or different users. The standard defines three levels of channel as shown in Figure 8.4: physical channels on which are mapped transport channels, each of which can carry one or more logical channels.

Logical channels are multiplexed onto transport channels by the MAC layer. Transport channels are multiplexed in the physical layer, but the mapping to physical channels is very much dependent on the configuration and type of channel. In some cases, a physical channel may carry several transport channels, and in other cases, a transport channel may be split across multiple physical channels. Channels can be divided into common channels and dedicated channels. Common channels are broadcast to all UEs in the downlink or shared by all UEs in the uplink [physical random access channel (PRACH)], in which case an access protocol is required to allow one UE to transmit at a time. DCHs are intended to be received or transmitted by a single UE, and no contention for access should occur. In Figure 8.4, the common channels are shaded lighter and the dedicated ones darker. Some channels are single instance only, but those shown with a shadowed image behind can have multiple instances. Transmission from the physical channels is divided into frames, each being 10 ms long. A frame is subdivided into five subframes of 2 ms, and each subframe into three slots of two thirds of a millisecond each, giving 15 slots per frame. The main significance of a slot is that it contains a single power control command, giving the capacity to send 1500 power control commands per second. This very fast rate of power control allows the network to ramp the

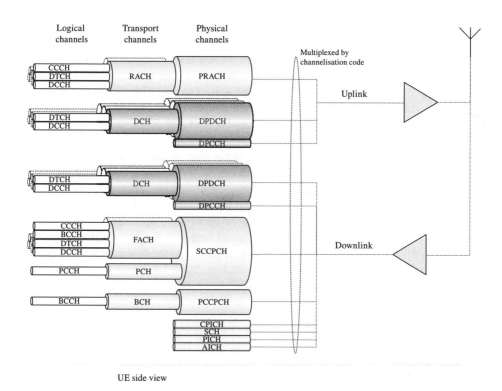

Figure 8.4 Organization of logical, transport and physical channels for the UTRAN (Release 99)

power up or down to cope with sudden signal path changes, for example when a user comes into line-of-sight with the basestation from behind a building, with minimum effect on other users, but it also presents a significant challenge to the UE. The frame and slot format shown in Figure 8.5 represents the generic format used for all channels, but within this, each

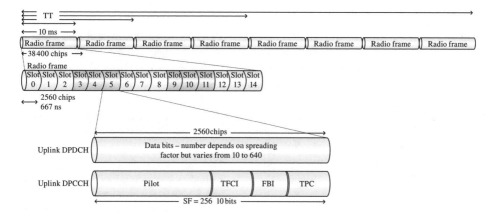

Figure 8.5 Generic radio frame format

channel has its own specific structure. TS 25.211 details each specific frame format, but to complete the picture, the formats for the dedicated physical data channel (DPDCH) and dedicated physical control channel (DPCCH) are shown here.

The DPDCH format is very straightforward, carrying only the user data bits. The DPCCH carries only control bits associated with the DPDCH. The main benefit of this separation is that higher data rates can be achieved by adding more DPDCHs, but still with only one DPCCH needed to carry their associated control. The DPCCH fields are as follows:

- Pilot: This carries a known bit pattern which, depending on slot format, can vary between 3 and 8 bits in length. This helps to maintain timing and frame synchronization and can also be used to adjust the receiver's equalizer.
- TFCI: The transport format combination indicator indicates which transport format (TF) combination has been selected and allows the receiver to demultiplex the transport channels carried by this physical channel. This is explained further in the next section.
- FBI: The feedback indicator is used to provide feedback to the basestation for controlling transmit diversity when it is operating in closed loop mode.
- TPC: This field contains the transmit power control commands to allow the UE to instruct the NodeB transmitter to raise or lower its power level. Two commands are provided: increase or decrease power, so if the power is to be held constant the commands need to alternate between up and down. The transmitter operates a filtering algorithm on the TPC pattern which prevents the power from oscillating and prevents corrupted commands from destabilizing the system. NodeB has a similar command mechanism operating on the UE transmitter.

8.1.5 Physical Channels

The various physical channels are summarized in Table 8.1.

Table 8.1   UTRAN physical channels

Acronym	Direction	Name	Description
Common channels			
PRACH	Uplink	Physical random access channel	UEs may access this channel at any time to initiate a connection with the network
P-CCPCH	Downlink	Primary common control physical channel	Carries the system broadcast information
S-CCPCH	Downlink	Secondary common control physical channel	Carries both signalling and data and can be read by all UEs in the cell
CPICH	Downlink	Common pilot channel	There are two of these, the primary and secondary common pilot channels (P-CPICH and S-CPICH), which provide phase references for the UE to accurately lock to the network timing. The P-CPICH is the

Table 8.1   (*continued*)

Acronym	Direction	Name	Description
			main reference and serves to allow the UE to decode the synchronization channels and can also be used to estimate the radio link quality and characteristics for channel equalization and for measurements
SCH	Downlink	Synchronization channel	There are also two of these, the primary and secondary synchronization channels (P-SCH and S-SCH), which provide frame and slot synchronization to the UE
PICH	Downlink	Paging indicator channel	This carries indicators that the UE can periodically monitor to determine whether it is being paged
AICH	Downlink	Acquisition indicator channel	This is used as part of the RACH protocol used by the UE to initiate a connection to the network. This channel is used to arbitrate contention on the PRACH
Dedicated channels			
DPDCH	Both	Dedicated physical data channel	Used to carry signalling and data between a UE and the network. Each DPDCH is dedicated to the use of one UE, and therefore its transmission power can be tailored according to factors such as how far away the UE is and how much interference it is experiencing
DPCCH	Both	Dedicated physical control channel	This is matched to a set of DPDCHs and carries format control and power control information

8.1.6 Transport Blocks and TFs

In a system where the transmitter power is rapidly controlled to balance the data rate given to the user with the interference he generates to other users, the data rate also needs to be rapidly varied. At the risk of over-simplifying, if the radio conditions are such that there is less available SNR, then the channel capacity will also reduce. In Figure 8.6 two systems are shown, one where the data rate is kept constant and one where it is varied dynamically. The line superimposed on the transport blocks indicates the real available channel capacity due to the SNR target. In the specifications, this is referred to as the signal-to-interference ratio (SIR) as the main component of the noise is coming from interference generated by other users. In this case, the SIR is continually changing, and this can be for a number of reasons: for example, more users are trying to access the cell or the UE has moved further from Node B. If the data rate remains constant, there comes a point where the SIR is too low to support the reception of the data, and most likely all the blocks transmitted at the same time as the diagonally shaded ones will be lost. Such a system would be quite inefficient.

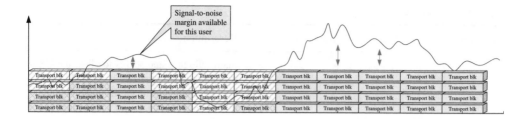

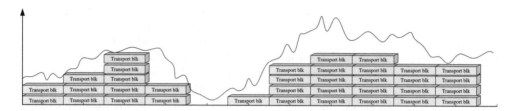

Figure 8.6   Varying data rate to cope with varying radio conditions

Dynamic rate control allows the transmitter to profile how much data it transmits according to the available SIR. In the UTRAN, this is achieved by grouping the bits for transmission into transport blocks and then allowing the transmitter to choose how many transport blocks to transmit within each TTI. This selection is complicated, however, by the need to be able to prioritize some data streams against others in order to provide different qualities of service on different links. The selection has to be communicated from the transmitter to the receiver so that it can work out what it has actually received. This is the task of the TFCI bits shown in Figure 8.7. Rather than list the number of blocks and their sizes for each transport channel, the method chosen is to put these into a table. The content of the table is shared between the transmitter and the receiver through upper layer signalling. The selection made can then be indicated at 'run time' by sending an index to the table. The table is known as the transport format combination set (TFCS), and hence, the index sent is the transport format combination index (TFCI).

The 3GPP standard provides a lot of flexibility to vary the factors effecting capacity, such as transmitter power, both for total and for individual channels, coding level, SF and so on, in order to optimize the capacity towards a variety of different network goals. One of the key advantages though is that the boundaries of speed or numbers of users can be made quite 'soft'. In GSM, the capacity in a cell was fixed by the number of timeslots and could only be increased by adding frequencies, placing hard limits on the system. For the WCDMA system, the performance for individual users can be reduced allowing more users onto the system. The reduction for individual users is statistical and gradual, and the network operator is able to spot areas where traffic levels are increasing and add more capacity before the level of service becomes a problem.

To support this flexibility, the data rate of each user needs to be continually adapted. This is done by dividing the data into transport blocks and then varying the number of transport blocks that are transmitted within a period of time. This transmission time is itself variable depending on the needs of the connection, as it is set by the period over which data is interleaved. As noted earlier in this chapter, interleaving provides the capability to

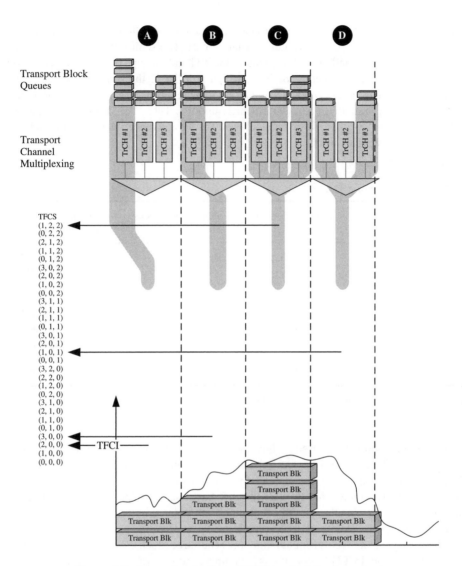

Figure 8.7 Dynamic selection of TFCI to adapt the channel rate

distribute bursts of errors more evenly over time and thus increases the chance of being able to successfully correct erroneous data. The longer the interleaving period, the better the system can correct larger error bursts, but the more latency the system suffers. Certain types of traffic need lower latencies and can tolerate dropped frames, whereas other applications need low frame loss but are not sensitive to latency. The interleave time is configurable by the network and can be set to 10, 20, 40 or 80 ms. This is known as the transmission time interval or TTI. Each transport channel has a set of different formats that can be selected (the transport format set or TFS). A transport channel can multiplex several logical channels, and each logical channel can have differently sized transport blocks, but within a TTI, different sizes cannot be mixed on one transport channel. The configuration of the transport channel is separated into a semistatic part and a dynamic part. The semistatic part can only change through upper layer signalling, in this includes the TTI and the type of error protection coding used. The dynamic part contains the TFS, which is a list of the supported transport block sizes and how many blocks of each size can be transmitted in a TTI. This list provides the granularity of data rate that particular channel supports. Table 8.2 gives an example of the TFS, based on the 384 kbps traffic channel used for the reference interactive or background packet-switched services bearer.

Table 8.2 Example of a TFS

TTI	TB size	Transport format	Number of blocks	Instantaneous data rate[a] (bits per second)
		TF0	0	0
		TF1	1	16 000
		TF2	2	32 000
		TF3	4	64 000
20 ms	336	TF4	8	128 000
		TF5	12	192 000
		TF6	16	256 000
		TF7	20	320 000
		TF8	24	384 000

[a] Adjusted for layer 2 headers (16 bits per TB).

An example of how this is used is shown in Figure 8.7. The example shows three transport channels multiplexed onto a physical channel. For simplicity, we will assume that all the block sizes are the same and that the TTI on each transport channel is the same. At point A, the transmitter determines that to meet the SIR target set by upper layer signalling, it can transmit two blocks. The three transport channels are ordered in terms of priority, and TrCH #1 has blocks queued ready, so those blocks are taken from its queue. In TTI B, the noise level has dropped, and the transmitter can now send three blocks. Again they are taken from the TrCH #1 queue. In TTI C, conditions have improved yet again, and the channel can work at its maximum of five blocks. This time, TrCH #1 can only supply one of them, so blocks are taken from TrCH #2. However, after emptying its queue there is still capacity, so the remaining space is filled by blocks from TrCH #3. At TTI D, a further block has arrived on TrCH #1, and this is taken. TrCH #2 has no pending blocks, so one is taken from TrCH #3. In each TTI, the receiver needs to know the quantities of blocks in each transport channel to correctly demultiplex them. A table is shown on the left-hand side, the TFCS. This table lists all the possible combinations of TFs across the three transport channels and is configured by upper layers. It is sent using RRC signalling and is known to both sides. In our simple case, the numbers refer to the number of transport blocks carried on each transport channel, with the first digit referring to TrCH #1, the second to TrCH #2 and the third to TrCH #3. As the table is common on both sides, all the transmitter needs to do is to send an index to the table, showing the element used for this transmission, and this is the TFCI.

8.1.7 Modulation

There is a consequence of being able to dynamically vary the data rate of a physical channel in this way. To see this, we need to look further at the way data are modulated onto the carrier. WCDMA uses quadrature modulation both in the uplink and in the downlink. In the uplink, this is mainly limited to a constant amplitude scheme, QPSK, where each symbol can have one of four states and hence encode two bits. The choice of a constant amplitude modulation has a number of benefits for the mobile terminal. In particular, it allows the use of efficient but nonlinear power amplifiers, where amplifying signals of varying amplitude forces the amplifier to operate at least for part of the time in its nonlinear region. This distorts the signal, spreading the energy outside of the useful bandwidth and making it difficult to stay within the constraints for out-of-band power set by the standard, and it also makes the amplifier inefficient, increasing the current drain on the battery and reducing the talk time. That said, as amplifier technology advances and more demanding applications emerge, there are niche areas where the reduced efficiency can be traded for the increased data rate. In Release 7, 3GPP is adding an optional 16-point QAM scheme as part of the adaptive modulation for the E-UL. For the downlink, high efficiency is not so important as, apart from power failure conditions, the Node B transmitter is generally driven from mains power. Node Bs can use less efficient but more linear amplifiers and thus support higher order modulation. In addition to QPSK, 16-point QAM and (from Release 7) 64-point QAM schemes are used on the high-speed downlink traffic channel.

Figure 8.8 shows an example of the spreading, modulation and scrambling for an uplink-dedicated physical channel. Multiple DPDCHs can be combined to increase the rate, and associated with them is a single DPCCH carrying the control information. The DPCCH has a constant SF of 256, which provides considerable processing gain. The SFs on the other channels vary according to the TFCIs selected. If the channels are using lower SFs, they will need a higher SIR to achieve a similar bit rate. In this case, if the channels are simply summed, the DPCCH (and other DPDCHs with higher SFs) will be transmitted with unnecessary extra power, increasing the noise in the system. This is prevented by associating gain factors β_d and β_c with each TFCI. Each physical channel is scaled by the appropriate gain factor after spreading, so while remaining within the overall power allowed to the UE, the way the power is shared between the channels can be optimized.

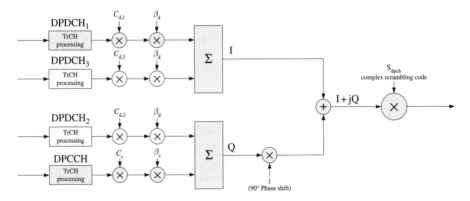

Figure 8.8 Spreading, modulation and scrambling for the uplink-DCH

8.1.8 Reference Measurement Channels

One of the consequences of the 'tool-box' approach chosen by 3GPP is that the system has too many independent parameters for exhaustive testing to be practical. The number of possible radio bearer configurations is very high, and radio bearers can be combined together in arbitrary combinations. Based on the old adage that if it is not tested, it will not work, infrastructure developers need configurations they can rely on having been tested. This makes the bearers chosen by the conformance test writers very important. As the conformance tests are widely used and in practice every UE will have been tested against them by the time they reach consumers, the configurations chosen for these tests represent a set of stable, well-known reference points. As discussed in Chapter 5, a substantial set of reference bearers are defined in TS 34.108 representing real network configurations. These bearers are extensively tested during signalling tests. However, the RF tests have slightly different requirements, and a set of bearer configurations which are slightly simplified forms of the equivalent normal reference bearers have been defined to support them. The reference measurement channels (RMCs) are defined in Annex A of TS 25.101 and repeated in Annex C of TS 34.121-1, and configurations covering a number of channel data rates are defined as follows:

- 12.2 kbps
- 64 kbps
- 144 kbps
- 384 kbps
- 768 kbps (uplink only).

Table 8.3 shows the parameter set for the 12.2-kbps RMC. There are two principal differences between this and the standard 12.2-kbps speech bearer from the reference set. First, the

Table 8.3 Definition of the 12.2 kbps reference measurement channel (TS 25.101)

Layer	Parameter	Channel	
		Traffic	Signalling
RLC	Logical channel type	Dedicated traffic channel (DTCH)	Dedicated control channel (DCCH)
	RLC mode	TM	UM/AM
	Payload sizes (bits)	244	88/80
	Max data rate (bps)	12 200	2200/2000
	RLC protocol data unit (PDU) header size (bits)	0	8/16
MAC	MAC header size (bits)	0	4
	MAC multiplexing	No	Yes
Layer 1	TrCH type	DCH	DCH
	Transport channel identity	UL: 1, DL: 6	UL: 5, DL: 10
	TB sizes (bits)	244	100
	TFS		
	TF0 (bits)	0×244	0×100
	TF1 (bits)	1×244	1×100

Table 8.3 *(continued)*

Layer	Parameter	Channel	
		Traffic	Signalling
	TTI (ms)	20	40
	Coding type	Convolution coding	Convolution coding
	Coding rate	1/3	1/3
	CRC size (bits)	16	12
	Maximum number of bits/TTI after channel coding	804	360
	Uplink: maximum number of bits/ radio frame before rate matching	402	90
	RM attribute	256	256

Note: CRC size: cyclic redundancy check codeword size; TTI: transmission time interval; TrCH: transport channel; RM attribute: rate matching attribute.

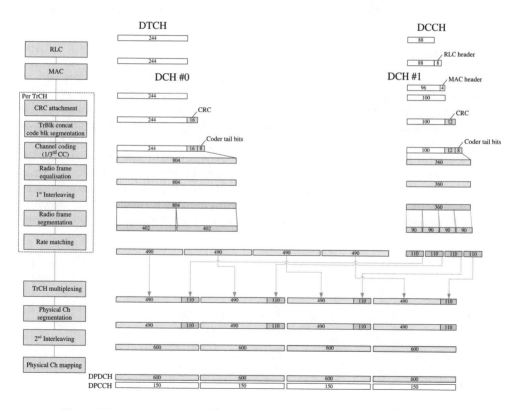

Figure 8.9 Layer 2 and transport channel processing flow for the 12.2 kbps RMC

speech bearer contains three multiplexed traffic flows, each on a different transport channel. The speech traffic from the voice coder is split into three different classes depending on the effect that corruption or loss of the data has on the quality of the decoded speech signal. The bits with the greatest effect are carried with the highest level of forward error correction coding. For the RF tests, this would complicate testing, so only one transport channel is used with the transport block size an aggregate of the three speech transport blocks. The second difference is that in the normal bearer the speech bearer is multiplexed with a low rate 3.4-kbps signalling channel split into four subchannels. This is simplified into a single channel at around 2 kbps, which carries dummy blocks during the RF measurements in order to simulate the conditions of simultaneous traffic and signalling.

The RMC definitions also include a useful illustration of how the channels are built up through the transport channel processing chain. In Figure 8.9, this is reproduced for the uplink direction and shown in context of the processing chain from TS 25.212.

8.2 Transmitter Testing

In general, transmitter tests are carried out by getting the UE into a state where it is continuously transmitting a signal that can then be accurately measured. The two key pieces of equipment needed to do this are the SS and the signal analyser. The SS provides the network side of the signalling necessary to get the UE into a transmitting state, and the signal analyser makes the measurements. A basic circuit diagram is shown in Figure 8.12.

8.2.1 Methods of Testing

Figure 8.10 shows a simplified diagram of a heterodyne transmitter. During development of the RF and IF stages, if direct conversion is not used, it can be useful to inject WCDMA signals either at digital I and Q, or at analogue I and Q, using a vector signal generator with WCDMA modulation capability. A signal analyser can then be used to look at the output of the RF stage by connecting at the antenna connection (Figure 8.11). Once the receiver has been tested and integrated with the transmitter, then the VSG can be used to generate a modulated RF signal into the receiver. This can be looped internally and the output measured with the signal analyser. However, the more complex testing begins during integration of the complete UE. Because of the sensitivity of RF designs to layout and parasitic coupling to surrounding circuitry, complete characterization of the design has to wait until UE is at form factor (i.e. layout, casings

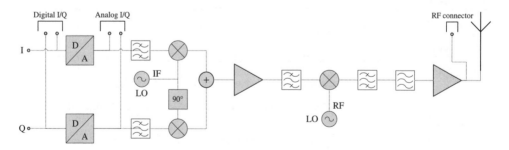

Figure 8.10   Simplified diagram of a heterodyne transmitter

Vector signal generator   Signal analyser

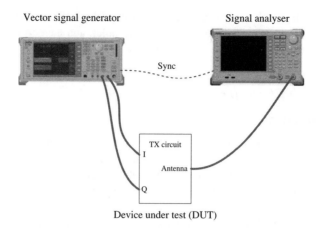

Device under test (DUT)

Figure 8.11 Configuration for testing the UE transmitter in isolation

and so on are representative of the commercial product) and the surrounding components, such as the baseband devices, processors, displays and so on, are in place and operating. At this point, RF testing usually takes place using special test modes or test software loaded into the baseband or through using the protocol stack to initiate a connection of some type.

Two common test environments are shown in Figure 8.12. In the first configuration (a), a signalling tester or SS is used to bring the UE into a connected state and to provide a stream of data on the downlink. The downlink data is looped back in the UE using one of the test loops mandated by the standard and transmitted on the uplink, which is monitored by the signal analyser. The second configuration (b) uses a manufacturing tester or radio communication analyser (RCA). This is a single instrument containing a real-time modulator and demodulator to provide the signalling and continuous data stream, but it also includes a signal analyser. The RCA has the advantages of being a single instrument and usually

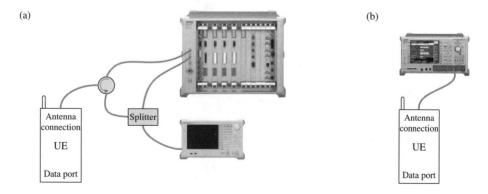

Figure 8.12 Configurations for testing the UE transmitter as part of the whole UE. (a) Testing with a SS and a signal analyser; (b) testing with a manufacturing tester

supports a variety of cellular radio standards in terms of both the necessary signalling and the preconfigured measurements. This makes it very quick and easy to use. However, it is usually not as accurate as a dedicated signal analyser.

8.2.2 Transmitter Characteristics

8.2.2.1 Transmitter Power

In Section 8.1, we have seen that noise from other users or neighbouring cells is one of the key sensitivities of a WCDMA network. While we are accepting a voluntary reduction in the channel SIR by making the signals from other users appear as uncorrelated noise, if they use more power than strictly necessary to achieve their desired quality of service, then this unfairly reduces the capacity available for other users. To maximize the capacity, we need to keep the signal power of each user as low as possible while still being able to decode the user's signal with an acceptably low error rate. This translates into accurately and continually controlling the power of each terminal's transmitter.

Power control is split into two parts: inner loop and outer loop. For the uplink, the radio network controller (RNC) manages the outer loop and Node B the inner loop (Figure 8.13). In the basic case of a single radio link, the RNC will set an initial power for the connection and a target SIR. The Node B estimates the SIR of the radio link and compares this against the target from the RNC. If it estimates that the UE is using more power than necessary to meet the SIR target, it will set the TPC bits to zero to command the UE to reduce power, and if

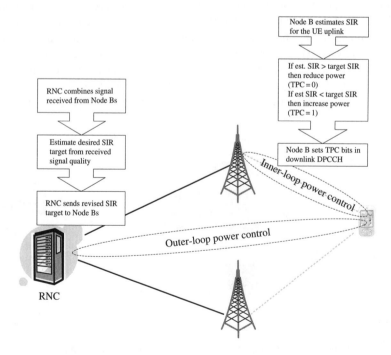

Figure 8.13   Inner- and outer-loop power control in the UTRAN

the UE is using too little power, then it sets the bits to one to command an increase. In this process, the UE has no direct knowledge of the SIR at the Node B's receiver, so it must behave as a slave to Node B. The power commands are relative, indicating a step up or down from the current level. The step size is variable and is configured for the link by the signalling. Of course, the power cannot rise or fall indefinitely, and limits on the upper and lower bounds are set. The upper bound, the maximum power, is set by the UE class and can also be set below the maximum capability of the UE through signalling when the link is established. The lower bound is defined in TS 25.101, which states that the UE must be able to reduce its power to at least –56 dBm.

There are two important points to note when measuring power during testing:

1. Most of the power measurements should be made after filtering with a root raised cosine (RRC filter) with a bandwidth set to the chip rate (3.84 MHz) and a roll-off (α) of 0.22. This removes power that is out of band as far as the Node B receiver is concerned and will typically give a value slightly lower (about 0.25 dB) than the unfiltered mean power. This filter is normally available within the measuring equipment and can be switched in or out.
2. In some cases, the conformance requirements in the test cases differ slightly from the specifications in TS 25.101. The reason is that the test specifications allow some amount of error or uncertainty in the test equipment and this is factored into the test requirements.

A typical test sequence for testing the UE's output power is illustrated in Figure 8.14. The procedure is as follows:

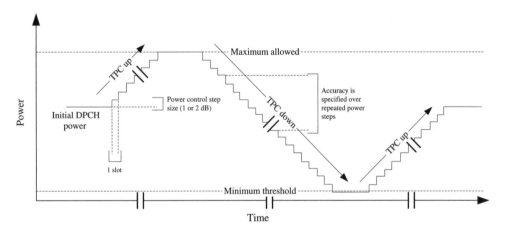

Figure 8.14   The operation of TPC commands

- The SS is configured to transmit a cell suitable for the UE to camp on to, and the UE is allowed to register.
- The UE is paged for a speech call and an RRC connection is established.
- If necessary, authentication and any security procedures are run.
- The UE is sent an Activate RB Test Loop command, which forces the UE to accept the call regardless of any NAS interaction but does not at this stage close any of the test loops.

- A Radio Bearer Setup command is sent, establishing one of the RMC channels.
- The test loop is closed; for this type of test, test loop 1 should be used.
- The signal analyser can be triggered to start measuring once power is detected on the uplink DPCH.
- The TPC bits on the downlink DPCCH are set to command the power continuously up long enough for the UE to reach its maximum power level, plus a little more to ensure that the power is clipped at that level.
- The TPC bits are now set to command the power continuously down long enough for the UE to reach its minimum power level.
- The power can be commanded back up again to a mid-value and then alternated to make sure the UE holds a steady power level.

This test would be rerun with different power step sizes and different network settings for the maximum allowed power. It can also be rerun with different power offsets between the DPDCH and the DPCCH and with different initial power settings. The resulting power versus time trace of such a test will allow the measurement of a number of values required in conformance tests, such as maximum output power and inner-loop power control, but in reality, the ability to do all these measurements on a single run will depend on the capabilities of the signal analyser, specifically the amount of memory it has for signal storage. The test may need to be broken into parts or rerun with different trigger conditions for different measurements. In the conformance tests, these measurements are split over several tests.

The requirements for the UE are shown in Tables 8.4–8.6.

Table 8.4 Maximum power classes

Power class	Maximum power (dBm)	Tolerance (dB)	
1	+33	+1.7	−3.7
2	+27	+1.7	−3.7
3	+24	+1.7	−3.7
3bis	+23	+2.7	−2.7
4	+21	+2.7	−2.7

Table 8.5 Applicability of power class in each frequency band

Operating band	Power class 1	Power class 2	Power class 3	Power class 3bis	Power class 4
Band I	✓	✓	✓		✓
Band II			✓		✓
Band III			✓		✓
Band IV			✓		✓
Band V			✓		✓
Band VI			✓		✓
Band VII			✓	✓	✓
Band VIII			✓	✓	✓
Band IX			✓		✓
Band X			✓		✓

Table 8.6 Power control step sizes

Step size (dB)	TPC command	Lower limit (dB)	Nominal (dB)	Upper limit (dB)
1	Up	+0.5	+1	+1.5
	None	−0.5	0	+0.5
	Down	−0.5	−1	−1.5
2	Up	+1.0	+2	+3.0
	None	−0.5	0	+0.5
	Down	−1.0	−2	−3.0

8.2.2.2 Open-Loop Power Control

The inner-loop power control mechanism is very effective for controlling the power on a DCH, but it requires a starting point as the commands are relative. It also cannot be used on common channels as these are shared by many UEs and it would be difficult to send each UE the necessary power commands. In addition, UEs can access the common channel in the uplink at any time using the random access protocol, so a mechanism is needed to ensure that power is effectively managed without direct feedback. This is the task of open-loop power control, and this is applied to the uplink DPCH and the PRACH physical channel at start-up. The initial power setting is determined dynamically by the UE based on a starting point signalled by the network. When the UE wants to access the PRACH or start transmitting the DPCH, it measures the received power on the primary common pilot channel (P-CPICH). This is a code domain power as the CPICH is code multiplexed with the other downlink channels (RSCP or received signal code power). The power at which P-CPICH is transmitted is signalled from the network in the system information, so from its measurement, the UE can estimate the path loss to Node B. This is not very accurate as the uplink and downlink are normally separated by a significant frequency gap and will be affected differently by fading. However, it gives an acceptable starting position. The estimated path loss and the signalled starting power level are used in an equation to set the transmit power accounting for the distance between the UE and the Node B. In the case of the PRACH, the UE transmits a preamble at this initial setting. If it receives no response, it assumes that the link conditions are too bad for successful reception and steps up the power according to a step size signalled from the network (Figure 8.15).

A test routine for open-loop power might look as follows:

- The SS is configured to transmit a cell suitable for the UE to camp on to. The power of the P-CPICH, and a signalled parameter called DPCCH power offset, are set to values calculated to simulate a specific path loss such that the initial power is less than the maximum allowed.
- The UE will attempt to transmit on the PRACH, and when power is detected, the signal analyser is triggered to start measuring.
- The SS is configured not to respond to RACH requests.
- After the RACH attempts finish, the cell is reconfigured (or another cell set-up) suitable for the UE to camp on to.
- This time the SS responds with an acknowledgement and the UE transmits the RRC connection request message.

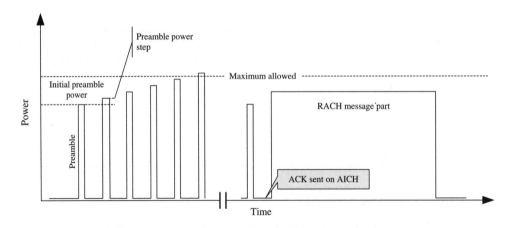

Figure 8.15 RACH preamble power step and message part power

Table 8.7 shows the nominal expected output power levels for the initial preamble, with the SS P-CPICH power set to simulate low, mid and high values of path loss.

Table 8.7 Expected output power levels for the initial RACH preamble (band I UE)

Parameter		Level	
Primary CPICH RSCP	–28.9 dBm	–69.9 dBm	–109.9 dBm
I_{or}	–25.0 dBm/3.84 MHz	–65.7 dBm/3.84 MHz	–106 dBm/3.84 MHz
Primary CPICH transmit power (signalled)	+19 dBm	+28 dBm	+19 dBm
Expected UE PRACH preamble initial power	–37.1 dBm	–13.4 dBm	+8.9 dBm

Note: Allowed tolerance is ±9 dB for normal conditions and ±12 dB over an extended temperature range (extreme conditions).

8.2.2.3 Transmit On/Off Mask and Transmit Off Power

The above test can also be used to measure the transmit on/off mask, which is basically the time taken for the power to switch on at the beginning of a transmission and off at the end, and to measure the off power, which is a level below which the UE must drop when it is not supposed to be transmitting. This is shown in Figure 8.16.

8.2.2.4 Power Control During Changing TFC

As we have seen in Section 8.1, the data rate on the channel can change dynamically through changing the TFCI. This change will usually be accompanied by a change in transmitted power.

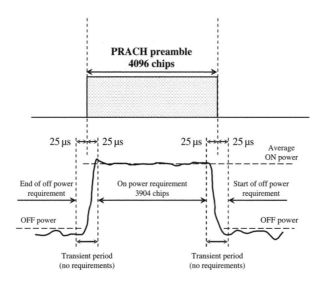

Figure 8.16   RACH preamble power profile (reproduced by permission of ETSI)

Requirements exist for both the transition time between the two levels and the accuracy of the output levels. Getting the UE to select different TFCs, however, is a little more complicated. TFC selection is done autonomously by the UE and can involve many factors, one of which is the amount of data buffered in the UE. The 12.2-kbps RMC is a fairly simple case as there are only two TFCs, one where the UE transmits one block on each channel and one where it does not transmit anything. The UE can easily be forced into selecting the 'off' TFC, as in test loop mode, the data is supplied from the SS through the downlink. If the SS stops sending data, then once all the data sent has been looped, the UE will have to select this TFC. This method is used in the conformance tests. For more extensive testing, particularly using configurations with a greater range of TFCs, the SS needs to send carefully measured amounts of data that will leave enough transport blocks in the UE to prompt the selection of the desired TFC. This process is a little unreliable, as the instantaneous buffer status will depend on the processing time in the UE and the coincidence of TTI boundaries, so the test may need to be repeated until the TFC is observed. The test can be made more reliable by sending larger amounts of data and using the RRC Transport Format Combination Control procedure to limit the TFC selection so that a higher rate one cannot be used. Figure 8.17 shows the time and power mask for TFC changes. The power accuracies depend on the step size, varying from ±0.5 dB for a 1-dB step to ±6 dB for a 21-dB step. The full table can be found in TS 25.101. A signal analyser which supports code domain power measurement should be used for this test, as the power change can also be accompanied by a change in the power offset between the DPDCH and the DPCCH (β_d and β_c). These channels are carried on different spreading codes, and their relative powers can be checked by looking at the code domain power of each.

8.2.2.5 Power Control During Loss of Synchronization

When the UE loses the downlink signal, then it can no longer receive power control commands. In this case, two things should happen. First, the UE should hold its current

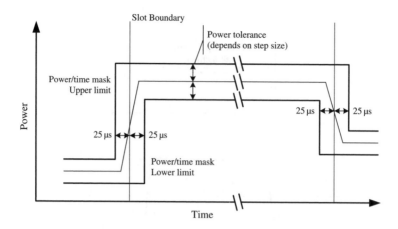

Figure 8.17 Time and power mask for a change in TFC (reproduced by permission of ETSI)

power level while it goes through the normal procedure (defined in TS 25.214) to find synchronization again. This procedure includes a time out after 160 ms, assuming that the UE does not see a strong enough signal from Node B within this time, at which point the UE should stop transmitting. The UE will continue trying to resynchronize until the upper layers time out, and if it does regain synchronization, then after 160 ms the UE can restart its uplink. The test procedure for this is as follows:

1. A call is set up using the 12.2-kbps RMC bearer and the uplink power can be set to maximum, so the downlink TPC bits can be set to one throughout the test. It is important that the upper layers do not time out during the measurement procedure. This can easily be done by setting extended values for T313 and N313, which are carried in SIB 1. The conformance test uses 15 s and 200, respectively.
2. The downlink power, however, needs to be carefully controlled to drop it below the threshold for keeping synchronization. In the conformance test, the power level is specified using the quantity E_c/I_{or}. This is actually a ratio of the energy in each chip divided by the power spectral density of the total signal output by the SS. In WCDMA, where the spectrum across the 3.84 MHz chip rate bandwidth is quite flat due to the use of the pseudorandom scrambling code, this approximates to the mean power times 3.84×10^6. The benefit of using these quantities is that they are much easier to relate to the actual gain settings and power level settings on individual channels in the SS.
3. The SS varies the output level of the DPCCH as shown in Figure 8.18. This needs to be done bearing in mind that the DPCCH E_c is a component of I_{or}, and when it is reduced either the calculation of the ratio needs to take this into account or another signal added (or increased) to compensate. The conformance tests allows a small amount of margin on all the E_c settings for all the critical thresholds (points A to E) to allow for SS output accuracy.
4. Measurements are taken after points C and F, and the criteria are that the power should be off (below −56 dBm) for the first and back on again for the second.

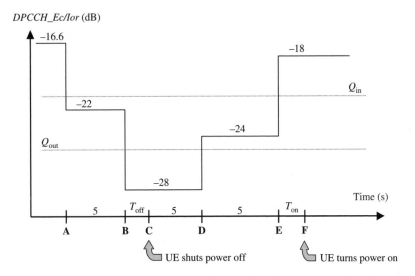

Figure 8.18 Power profile on downlink DPCCH for testing power control during loss of synchronization (reproduced by permission of ETSI)

8.2.2.6 Frequency Control

The carrier frequency is generated by multiplying up a lower frequency reference oscillator using a digital counter (often called a numerically controlled oscillator or NCO). This is necessary as the UE has to be able to set the carrier frequency under software control to tune into the assigned UTRA absolute radio frequency channel number (UARFCN). The actual frequency of operation for any UARFCN is a compromise of the closest that can be achieved based on the reference frequency and the discrete steps of the multiplying chain. The requirement on the UE from TS 25.101 is that the actual frequency should be within ± 0.1 parts per million of the nominal frequency and that it should not vary outside of this. The UE derives its transmit frequency depending on the frequency of the cell it camps on to, so in this case, the nominal frequency is defined as the Node B transmitter frequency less the defined uplink–downlink separation. This can be measured using the same set-up as for the power measurements and is carried out by setting up a call using an RMC in test loop mode, but with the SS signal level set to the receiver reference level. This is explained further in Section 8.3, but is the lowest level the receiver should operate reliably at, and provides the worst case conditions for the UE to lock to the Node B transmitter. The carrier oscillator will also need some time to settle to the required frequency, and 25 μs is allowed in the specifications. This test can also be repeated over a range of temperatures and battery voltages as these factors can also affect the stability and accuracy of the oscillator.

8.2.2.7 Transmitter Spectrum

Imperfections in the local oscillators, mixers, gain stages and filters of the RF chain all serve to generate signal power outside of the 5 MHz bandwidth used for useful data transmission.

These waste energy and thus reduce talk and standby times in the terminal, and they also create interference in neighbouring channels. Some level of imperfection is unavoidable, but limits are set on the resulting effects. Four key measures are used:

- Occupied bandwidth
- Spectrum emission mask
- ACLR
- Spurious emissions.

The test set-up used for power measurements can also be used to make these measurements. The general procedure is to set up a call using the 12.2-kbps RMC configuration and, allowing a little settling time, trigger the signal analyser once power is detected on the uplink DPCH.

Occupied bandwidth is defined as the bandwidth within which 99% of the transmitted power is contained, and the requirement is that this should not be greater than 5 MHz. This should be a built-in measurement on the signal analyser, as shown in Figure 8.19.

The spectrum emission mask sets a power versus frequency template around the transmitted bandwidth. The mask is specified both relative to the carrier power (dBc) and in absolute power terms, with the relative value applying when the carrier level is above the minimum 'off' value of –56 dBm. The measurement in the area closest to the carrier (2.5–3.5 MHz away) is made using a Gaussian filter of 30 kHz bandwidth, while in the further areas (beyond 3.5 MHz), the filter bandwidth is extended to 1 MHz. The carrier power, which is needed to get the relative measurement, is measured with a 3.84-MHz RRC filter as for the other

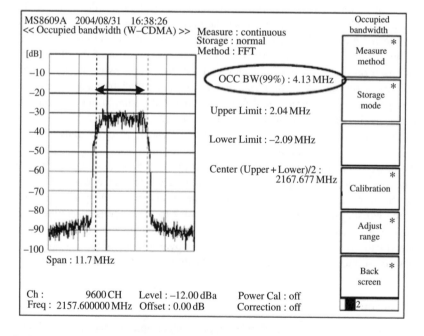

Figure 8.19 Typical plot of occupied bandwidth for a WCDMA transmitter (reproduced by permission of Anritsu)

power measurements. Again, high-end signal analysers which have built-in measurements for WCDMA will calculate and display the mask based on the measured carrier power and will apply the correct filtering. Table 8.8 shows the requirement from TS 25.101. The actual allowed values are relaxed slightly in the conformance tests to allow for test equipment measurement uncertainty.

Table 8.8 Spectrum emission mask for the UE (reproduced by permission of ETSI)

Δf in MHz[a]	Minimum requirement[b]		Additional requirements bands II, IV, V, X^c (dBm)	Measurement bandwidth[d]
	Relative requirement	Absolute requirement (dBm)		
2.5–3.5	$\left[-35-15\left(\dfrac{\Delta f}{\mathrm{MHz}}-2.5\right)\right]$ dBc	-71.1	-15	30 kHz[e]
3.5–7.5	$\left[-35-1\left(\dfrac{\Delta f}{\mathrm{MHz}}-3.5\right)\right]$ dBc	-55.8	-13	1 MHz[f]
7.5–8.5	$\left[-35-10\left(\dfrac{\Delta f}{\mathrm{MHz}}-7.5\right)\right]$ dBc	-55.8	-13	1 MHz[f]
8.5–12.5 MHz	-49 dBc	-55.8	-13	1 MHz[f]

[a] Δf is the separation between the carrier frequency and the centre of the measurement bandwidth.
[b] The minimum requirement is calculated from the relative requirement or the absolute requirement, whichever is the higher power.
[c] For operation in bands II, IV, V, X only, the minimum requirement is calculated from the minimum requirement calculated as in footnote b or the additional requirement for band II, whichever is the lower power.
[d] As a general rule, the resolution bandwidth of the measuring equipment should be equal to the measurement bandwidth. However, to improve measurement accuracy, sensitivity and efficiency, the resolution bandwidth may be smaller than the measurement bandwidth. When the resolution bandwidth is smaller than the measurement bandwidth, the result should be integrated over the measurement bandwidth to obtain the equivalent noise bandwidth of the measurement bandwidth.
[e] The first and last measurement position with a 30-kHz filter is at $\Delta f=2.515$ and 3.485 MHz.
[f] The first and last measurement position with a 1-MHz filter is at $\Delta f=4$ and 12 MHz.

The spectrum emission mask is aimed at picking up narrow peaks of power in the neighbouring channels, such as might be caused by other device clocks coupling into the modulated signal. The ACLR is closely related but measures the average power (RRC filtered) across the whole bandwidth of the channels next to and next but one to the assigned frequency. The requirement is that the channel at ± 5 MHz should be at least 33 dB below the assigned channel, and the channel at ± 10 MHz should be 43 dB below. The measurement of ACLR is shown in Figure 8.20.

The last of this set of measurements is for spurious emissions. These also come from real-world imperfections in the transmitter chain, such as nonlinearity in the amplifiers, parasitic coupling of other signals or images from the D/A converters, but in this case, the search window is across a very wide range of frequencies from 9 kHz up to 12.75 GHz. There

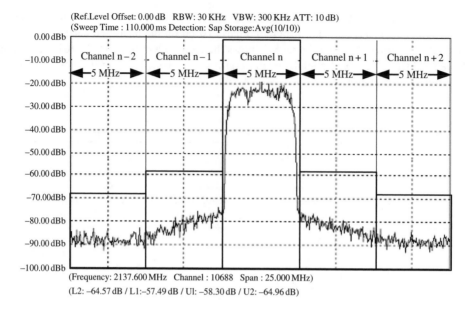

(Ref.Level Offset: 0.00 dB   RBW: 30 KHz   VBW: 300 KHz ATT: 10 dB)
(Sweep Time : 110.000 ms Detection: Sap Storage:Avg(10/10))

(Frequency: 2137.600 MHz   Channel : 10688   Span : 25.000 MHz)
(L2: –64.57 dB / L1:–57.49 dB / Ul: –58.30 dB / U2: –64.96 dB)

Figure 8.20   A typical plot of ACLR (reproduced by permission of Anritsu)

are two exceptions. The frequencies around the carrier are not checked as these are more stringently measured in the preceding tests, and there are some more onerous requirements across the other bands used for wireless systems. The general requirements are shown in Table 8.9, including the specific requirements for a band I UE, and Figure 8.21 shows how this applies to some of the other wireless systems.

Table 8.9   Maximum allowed levels of spurious emissions for a band I UE

Frequency bandwidth	Measurement bandwidth	Minimum requirement (dBm)
9 kHz $\leq f <$ 150 kHz	1 kHz	–36
150 kHz $\leq f <$ 30 MHz	10 kHz	–36
30 MHz $\leq f <$ 1000 MHz	100 kHz	–36
1 GHz $\leq f <$ 12.75 GHz	1 MHz	–30
Additional requirements for specific frequency ranges for a band I UE		
860 MHz $\leq f \leq$ 895 MHz	3.84 MHz	–60
921 MHz $\leq f <$ 925 MHz	100 kHz	–60[a]
925 MHz $\leq f \leq$ 935 MHz	100 kHz	–67[a]
	3.84 MHz	–60
935 MHz $< f \leq$ 960 MHz	100 kHz	–79
1805 MHz $\leq f \leq$ 880 MHz	100 kHz	–71[a]
1844.9 MHz $\leq f \leq$ 1879.9 MHz	3.84 MHz	–60

Table 8.9   *(continued)*

Frequency bandwidth	Measurement bandwidth	Minimum requirement (dBm)
1884.5 MHz < *f* < 1919.6 MHz	300 kHz	–41
2110 MHz ≤ *f* ≤ 2170 MHz	3.84 MHz	–60
2620 MHz ≤ *f* ≤ 2690 MHz	3.84 MHz	–60

ᵃ The measurements are made by sweeping through the frequency band in 200 kHz steps. As exceptions, up to five measurements are permitted for each UARFCN where the level is greater than that specified, but in all cases it should be below the overall level for that frequency range (i.e. –36 or –30 dBm as appropriate).

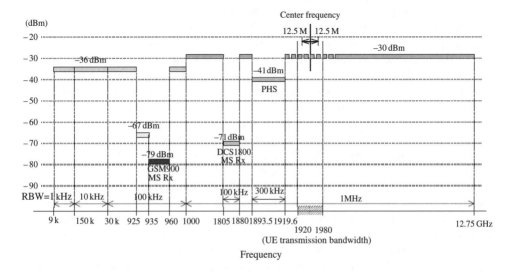

Figure 8.21   Permitted spurious emission levels across the radio spectrum (reproduced by permission of Anritsu)

8.2.2.8 Modulation

The uplink channels up to Release 6 use a phase-modulated signal with a four-point constellation, known as QPSK. Two enhancements are applied to increase the performance of the modulation scheme when used with the typical nonlinear amplifiers of a hand-held device, and these are the use of a RRC filter on the baseband signal and the use of a special class of complex scrambling codes that reduce the number of 180° phase transitions. These enhancements reduce the peak-to-average power ratio (PAR) and allow the power amplifier stage to work more efficiently.

The accuracy of the modulated signal generated by the transmitter has a direct effect on the performance of the receiver in the Node B. The receiver needs to make a decision as to which constellation point a received symbol maps to. With the addition of random noise, as the noise level increases, the probability that the receiver sees the symbol as closer to the wrong constellation point increases, and thus, the bit error rate (BER) increases. The noise level depends on both the externally added channel noise and the internally added noise from the transmitter itself. The main sources of internal noise come from phase noise added in the RRC modulation filter, accuracy, linearity and quantization in the D/A converters, imbalances between the I and Q paths, phase noise in the local oscillators and nonlinearity in the power amplifier. Three measures are used to qualify the modulation quality:

1. Error vector magnitude (EVM)
2. Code domain error
3. Phase discontinuity.

Measurement of these quantities requires the UE to transmit a continuous data signal, so the set-up, once again, is the same as for the power measurement tests, and the test procedure is the same as the previous measurements, requiring the establishment of a call using the 12.2-kbps RMC bearer in test loop mode 1. The analyser can be triggered once power is detected on the uplink DPCH, allowing 50 μs the signal to settle. For conformance tests, some of these measurements are carried out at maximum power, so inner-loop power control is enabled and the TPC bits set continuously to ramp the power to maximum. For conformance test measurements at less than maximum power, inner-loop power control is enabled and closed loop power control is used.

The EVM measures the difference between the ideal signal and the real transmitted signal (Figure 8.22). To get the ideal signal, the test set needs to demodulate back to the original I and Q data and then regenerate the signal as it should have been transmitted. This is called the reference signal and allows the test set to remove other artefacts such as clock mismatches and path losses by minimizing the EVM. The residual then represents the true error in the transmitted signal. As can be seen from the diagram, the error vector is an instantaneous measurement, so to be useful, it is averaged over a period of time (usually one slot), actually using the mean power, and is expressed as a percentage of the square root of the mean reference signal power. The conformance requirement is that the mean EVM should not be greater than 17.5%. For general testing, it is also useful to look at the peak EVM and the constellation plot itself. Most signal analysers will display a cumulative constellation plot where each received sample is plotted as a point. The distribution of constellation points will often give clues as to the nature of any transmitter problems.

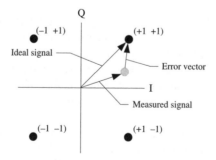

Figure 8.22   Error vector magnitude

The mean EVM gives a quality measure for the overall signal, and in the cases where only one or two channels are transmitted (PRACH or DPDCH/DPCCH), is the only measurement required. However, for multicode operation, the PAR of the signal is increased as the signal for each code is summed, and this places much stricter linearity requirements on the power amplifier. The overall measure may mask problems in individual code channels. Hence, an additional measurement is also made of the peak code domain error. In this measurement, the power of the error vector is measured in each code at a given SF and expressed as a ratio (in dB) to the mean power of the reference signal. The largest (peak) code domain error measurement should not exceed −15 dB. In the conformance test, the measurement is made at two different total output power levels: maximum and −18 dBm.

8.2.2.9 Transmitter Intermodulation

If the transmitter is operating in the presence of another strong signal at a close frequency, such as another UE nearby, the signal can couple back down the antenna into the power amplifier and create intermodulation products which are then amplified and transmitted back out. Although these signals are likely to be much smaller than the UE's own signal, they can create additional spurious signals. In testing, this is simulated by injecting an interference signal at 5 and 10 MHz offsets from the UE's carrier. The test set-up differs slightly from that used previously, in that a signal generator is included to provide the interferer, as shown in Figure 8.23.

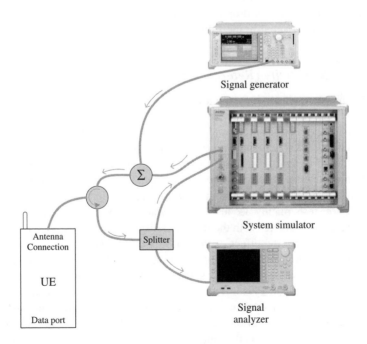

Figure 8.23   System for measuring transmitter intermodulation products

In this test, the signal generator is set up to transmit a continuous wave (CW) signal below the UE carrier power. A call is set up using the standard RMC bearer, and once the uplink DPCH signal is detected, the signal analyser can be triggered. The intermodulation products occur at offsets related to the difference between the two frequencies, both above and below the UE frequency. The mean power of the intermodulation products is measured through an RRC filter and compared against the filtered mean signal power. For the 3GPP conformance test, the CW signal is injected at −40 dBc and the requirements for maximum power of the intermodulation products are given in Table 8.10.

Table 8.10 Requirements for maximum power of the intermodulation products (reproduced by permission of ETSI)

CW signal frequency offset from transmitting carrier	5 MHz	10 MHz
Interference CW signal level	−40 dBc	
Intermodulation product	−31 dBc	−41 dBc

8.3 Receiver Characteristics

Testing the receiver is also important from the network's perspective. If the receiver is not sensitive enough, then Node B will need to use a higher signal level to achieve the same downlink quality of service, and this will increase the noise and reduce the capacity in neighbouring cells. This also has a bearing on UE quality, of course, as the network will not indefinitely provide more power, and poor receiver performance eventually shows up as a greater tendency to drop calls or drop out of service.

The areas to test are covered in the following sections.

8.3.1 Receiver Sensitivity (Reference Sensitivity Level)

Typically, the receiver sensitivity is measured in terms of BER at a given input signal to noise level. The reference level is the case where no noise is added to the downlink, so the receiver sees only the thermal noise added by its own input stage. In practical terms, this involves transmitting a known data pattern on a downlink channel at a specific power level (usually specified in terms of E_c/I_{or} for the channel). The data is demodulated by the UE, looped back, modulated and retransmitted to the SS. The received data is then compared against what was transmitted and the number of mismatched bits counted as a percentage of the total sent. These tests mainly use test loop 1, which loops data at the top of the RLC, but the RMC configuration operates with the RLC in transparent mode and with no MAC header. Effectively, we can ignore the effect of these layers on the data itself, so the loop appears to be at the top of the transport channel-processing chain. The data pattern sent is usually a pseudorandom binary sequence using a standard generator. Two are commonly used, and these are known as PN9 and PN15. They are quite widely used for BER testing because their pseudorandom nature gives them a flat spectrum across the communication bandwidth, and in many cases, this is important for properly testing demodulation and decoding. In WCDMA, this is not necessary as the data stream is already randomized by the scrambling sequence; however, they also have one other benefit. Being pseudorandom, they have strong autocorrelation properties, and therefore, the BER measurement instrument does not need to be synchronized with the transmission of data on the downlink. It should be able to gain synchronization by correlating its PN sequence against the received data.

A typical test configuration for receiver sensitivity testing is shown in Figure 8.24. The basic receiver sensitivity test only needs the SS and a BER tester; however, other receiver tests also require a signal generator for the injection of other signals, and the BER test functionality is often built in to the VSG.

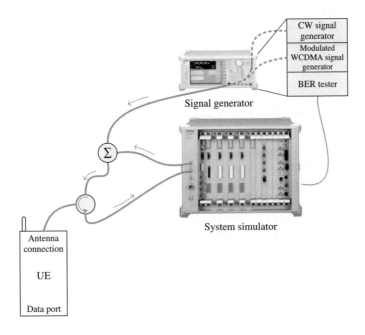

Figure 8.24   System for measuring receiver sensitivity

A general test procedure for a sensitivity test is as follows:

- A cell is set up suitable for the UE to camp on to, and the UE is allowed to register.
- The UE is paged for a speech call and an RRC connection is established.
- If necessary, authentication and any security procedures are run.
- The UE is sent an Activate RB Test Loop command, which forces the UE to accept the call regardless of any NAS interaction but does not at this stage close any of the test loops.
- A Radio Bearer Setup command is sent, establishing one of the RMC channels. The reference sensitivity tests in the conformance specification (TS 34.121) uses the 12.2-kbps RMC bearer, but for general testing, other bearers can be selected.
- The test loop is closed; for this type of test, test loop 1 should be used.
- The SS sets the TPC bits to ones continuously to ramp the UE uplink power to maximum.
- The SS starts to transmit a PN9 data sequence on the downlink channel under test (the DPCH for the reference sensitivity).
- The data is received and decoded on the uplink channel and the BER measurement is made.
- The measurement is continued until enough data samples have been taken to provide a meaningful result.

The sensitivity test can be run with two goals. The conformance test sets a specific power level and then specifies a maximum acceptable BER at that level. However, it can also be useful to set a specific BER level and then gradually reduce the power until the BER figure is reached. In either case, the number of samples to gather is statistically determined and depends on the confidence level required for the measurement. In general though, these tests are run for significant periods of time, and TS 34.121 provides some recommendations in Annex F for early termination of testing to reduce the test times.

The conformance requirement for the reference sensitivity level varies from band to band, but for band I, the receiver must show a BER of one in a thousand or less (0.001) with the DPCH E_c set to –117 dBm and the I_{or} set to –106.7 dBm measured at the UE antenna connection.

8.3.2 Maximum Input Level

The UE receiver should be able to support a large dynamic range, and problems with the automatic gain control will show up with high signal levels, such as when it is operating close to the basestation. Testing the maximum input level puts a strong signal into the UE receiver and measures the BER in the same way as the previous test. Again, the input signal power can be set and the BER measured, or the BER level set and the input signal power increased until the BER level is reached. The downlink power level is chosen to simulate a low path loss between a high power Node B and the UE. The conformance test sets the downlink signal I_{or} at –25.7 dBm/3.84 MHz at the UE antenna, with the DPCH E_c –19 dB below this. The BER level must not exceed 0.001.

8.3.3 Adjacent Channel Selectivity

This tests the ability of the UE to receive in the presence of a signal on a neighbouring frequency channel. The UE should band-limit the signal to the 5-MHz channel it is tuned to, but the filters will not remove all of the out-of-band energy from the adjacent channel, and this will reduce the sensitivity of the receiver. Once again, the sensitivity is measured with a BER test using the same procedure as before, but this time a WCDMA-modulated signal is added at a 5-MHz offset. This needs to look like a signal from another ULTRAN cell on an adjacent frequency channel, and in the conformance test, the signal is configured as shown in Table 8.11, with a typical configuration of common channels and with 16 channels of random data spread by OSVF codes simulating traffic from other users. This is known as orthogonal channel noise simulation (OCNS) and is also used in a number of RF tests discussed later in this section.

The conformance requirement was tightened up between Release 99, where the test is run at low power, and Release 5 onwards, where a higher power test is added. For the low-power case, the DPCH E_c and the I_{or} are set to the 14 dB above the reference sensitivity test levels for the band of operation being tested, and the adjacent channel interference power is set to –52 dBm. For the high-power case, the DPCH E_c and the I_{or} are set to the 41 dB above the reference sensitivity test levels and the adjacent channel power increased to –25 dBm. In both cases, the requirement is that the BER should not exceed 0.001.

Table 8.11 Configurations of channels on a neighbour cell when testing adjacent channel selectivity

Channel type	Spreading factor	Channelization code
P-CCPCH	256	1
SCH	256	–
P-CPICH	256	0
PICH	256	16
OCNS channels		
DPCH	128	2
DPCH	128	11
DPCH	128	17
DPCH	128	23
DPCH	128	31
DPCH	128	38
DPCH	128	47
DPCH	128	55
DPCH	128	62
DPCH	128	69
DPCH	128	78
DPCH	128	85
DPCH	128	94
DPCH	128	125
DPCH	128	113
DPCH	128	119

8.3.4 Blocking Characteristics

The blocking characteristics are basically an extension of the adjacent channel selectivity test, looking over a much broader set of frequencies than just the neighbouring channel. Three types of test are defined:

Type	Interferer	Purpose
In-band blocking	WCDMA-modulated interferer	This is to look for sensitivity to interference from other networks in the WCDMA band served by the UE, but outside of the neighbouring channels tested previously. This also includes a small 'guard area' of 15 MHz above and below the band.
Out-of-band blocking	CW interferer	This is for general interference anywhere in the spectrum outside of the operating band.
Narrow-band blocking	Gaussian minimum shift keying (GMSK)-modulated interferer	This tests for sensitivity to signals from second-generation cells in bands II, III, IV, V, VII and X where there are networks in close frequency proximity. The interferer is placed at a frequency of 2.7 or 2.8 MHz offset from the receiver channel.

The conformance requirements for the powers for the in-band and out-of-band blockers for a band I UE are shown in Figure 8.25. The test procedure follows that in the previous tests, but once the bearer is established, the appropriate interferer is swept through the area of spectrum in 1-MHz steps, with the BER measured at each step. Consequently, this test can take a considerable time to run. The BER should not exceed 0.001.

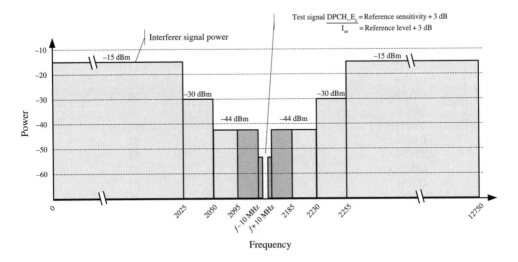

Figure 8.25 In-band and out-of-band blocking characteristics for a band I UE

8.3.5 Spurious Response

During the blocking characteristics test, there are likely to be some hot spots where, due to the shape of the filter characteristics, the attenuation of the interferer is not enough and the BER requirement cannot be met. Providing these occur in the out-of-band region, the UE is allowed a certain number of these hot spots (24 per test frequency in the case of a band I UE). These frequencies are called the spurious responses and are given the slightly relaxed requirement that the UE should not exceed the BER limit of 0.001 with a CW interferer of –44 dBm.

8.3.6 Intermodulation Characteristics

The adjacent channel, blocking and spurious tests check the performance of the receiver in the presence of signals at various other frequencies around the intended signal, but it is also possible for two nearby signals to generate intermodulation products that fall directly into the receive frequency. For example, the UMTS spectrum licenses are allocated in 5-MHz or multiples of 5-MHz chunks. It is likely that a UE on one network will be in the vicinity of other UEs on networks both 10 and 20 MHz offset, as operators' networks tend to overlap within a territory. These will intermodulate to produce a harmonic directly in the receiver band of the UE. The same can happen with bands colocated with 2G networks.

The configuration and procedure for this test is almost the same as the previous receiver tests, but in this case, two interference signals are required. This can be done with two signal generators and a three-way combiner, but more typically a dual-channel signal generator is used so that the test set-up does not need to be changed.

There are two conformance tests associated with this. One uses a CW interferer at 5 MHz offset and a WCDMA interferer at 10 MHz offset. They are both set to –46 dBm mean power, and the wanted signal is transmitted at 3 dB above the reference sensitivity level. The other uses a CW interferer and a narrow-band GMSK interferer set-up as shown in Table 8.12. In both tests, the requirement is that the BER should not exceed 0.001.

Table 8.12 Test parameters for narrow band intermodulation characteristics (reproduced by permission of ETSI)

Parameter	Unit	Bands II, IV, V and X		Bands III and VIII	
DPCH_E_c	dBm/3.84 MHz	Reference sensitivity level +10 dB		Reference sensitivity level +10 dB	
$\hat{I}_{or}$	dBm/3.84 MHz	Reference $\hat{I}_{or}$ level +10 dB		Reference $\hat{I}_{or}$ level +10 dB	
I_{ouw1}(CW)	dBm	–44		–43	
I_{ouw2}(GMSK)	dBm	–44		–43	
F_{uw1} (offset)	MHz	3.5	–3.5	3.6	–3.6
F_{uw2} (offset)	MHz	5.9	–5.9	6.0	–6.0
UE transmitted mean power	dBm	20 (for power class 3) 18 (for power class 4)			

Note: I_{ouw1} is the mean output power of the first interferer and F_{uw1} is its offset frequency from the required signal. I_{ouw2} is the mean output power of the second interferer and F_{uw2} is its offset frequency.

8.3.7 Spurious Emissions

This test checks that the signals within the receiver circuit do not couple back into the transmitter or antenna and cause excessive interference. Although this is a receiver test, as it is testing radio emissions, the test set-up is the same as for some of the transmitter tests as shown in Figure 8.12a, where a wideband signal analyser or spectrum analyser is used to scan for these emissions throughout the spectrum. The test sequence is also a little different to the other receiver tests. It needs to be done with the receiver active but with no normal activity in the transmitter. This is achieved by putting the UE into the RRC CELL_FACH state (see Section 10.2.2 for more details on RRC states), where the UE only transmits when it has data to send. The SS downlink signal is configured with an active S-CCPCH which can transmit dummy data on the forward access channel (FACH). The UE will attempt to decode the FACH as it is a shared channel and determine at the MAC level whether the data is directed to it. There is no need to close the test loop, but activating the test loop mode may still be useful to prevent interactions from the upper layers. The system information needs to be set such that the UE will not try to reselect to another cell during the test, and this is done by setting the relevant parameters ($S_{\text{intersearch}}$, $S_{\text{intrasearch}}$, Q_{qualmin} and Q_{rxlevmin}) to levels that prevent the UE from searching for alternative cells.

The test procedure will set up an RRC connection, bringing the UE into CELL_FACH state. Dummy data is transmitted on the FACH and the signal analyser triggered to sweep through the RF spectrum from 30 MHz up to 12.75 GHz. Up to 1 GHz a measurement filter of 100 kHz is used, and above this, the filter is set to 1 MHz. The conformance requirement is summarized in Table 8.13.

Table 8.13 General receiver spurious emission requirements

Frequency band	Measurement bandwidth	Maximum level (dBm)
30 MHz $\leq f <$ 1 GHz	100 kHz	−57
1 GHz $\leq f \leq$ 12.75 GHz	1 MHz	−47

In addition to these measurements, an additional test is made for spurious emissions within the operating band of the UE or within related 2G bands for some of the bands, where the requirement is increased. For example, for a band I UE, the spurious emissions within the band should not exceed −60 dBm.

8.4 Interactions with GSM Technology

In most regions, UMTS handsets are dual-mode, supporting both WCDMA and GSM operation. The operator has a degree of flexibility to roll-out WCDMA as an overlay on his existing network and can direct traffic onto either network according to his operational strategy. However, to support this the UE needs to be able to make measurements of both the WCDMA network and the GSM network and report these back to the Node B/RNC. Measuring within the WCDMA network is straightforward, as mostly the cells operate on a single 5-MHz frequency and are separated in the code domain. However, some networks have more than one 5-MHz allocation, and of course, the GSM cells are all on different frequencies. The WCDMA air interface link is continuous, so in normal operation when the UE is in connected mode, there are no convenient points, such as other user's timeslots, where it can retune its receiver to measure at a different frequency. This could be solved by having two receivers, but this increases the cost and power consumption of the terminal and so it is undesirable to have this as a mandatory requirement. Instead, the standard includes a mode of operation called compressed mode.

In compressed mode, gaps in the connection are created by switching off transmission during some of the slots in a frame. During the gap, the UE can briefly retune its receiver to a different frequency and make a measurement. The gaps can be created in one of two ways. A lower SF (half of the current SF) can be applied for that frame, allowing the same amount of data to be transmitted in half the time. This method is known as SF/2 and has the advantage that the user's data rate can be maintained but at the cost of a slight reduction in capacity. Alternatively, for TTIs with frames that contain a gap, a TFCI can be used that provides less data to the physical layer than the current SIR can support, allowing the physical layer to send all the data in a shorter time and hence create the gap. This method is known as higher layer scheduling and provides the opposite benefit, in that it maintains capacity at the cost of a slight reduction in the user's data rate. Within one frame, the maximum gap that can be created is seven slots, but with either method, larger gaps (up to 14 slots) can be created by positioning a gap at the end of one slot and at the start of the next. Compressed mode can be individually applied to the downlink as it is the UE receiver that is used to make the measurement, but in many cases, the receiver and transmitter share a common local oscillator, and it is useful for both uplink and downlink to stop at the same time. The measurements made during the compressed mode gaps are covered in more detail in Section 10.2.8, but compressed mode also places some power control requirements on the UE.

There are two considerations for power control. First, when SF reduction is used, the lower SF requires more power to achieve the same data rate. Second, where there is no downlink transmission during the gap, the UE will not receive any power commands. Once the transmission restarts, the UE must recover lost ground and quickly converge its power level to meet the SIR target. There are some special power control features that handle these considerations. As with many features, there are a number of options or modes, but a simplified view is shown in Figure 8.26. During the frames containing the gaps and compressed data, the UE power is stepped up by adding an offset to the SIR target. At the end of the gap, the UE has a choice of algorithms (directed by the network) to estimate where to set its power level for the first slot. Following that, a larger power step size is used for a short period, known as the recovery period.

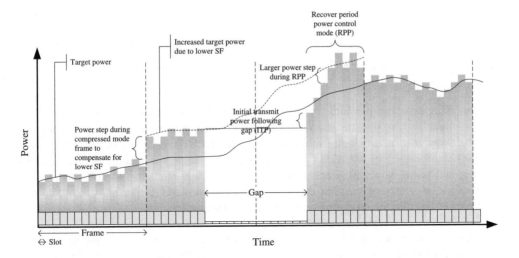

Figure 8.26 Compressed mode measurement gap with power control adjustments either side

The conformance test checks only SF/2 compressed mode and limited modes of the various options. For full details, see clause 5.7 of TS 34.121.

8.5 Performance Testing

So far, the RF tests have focussed on very specific characteristics of the receiver. The performance tests provide a broader test of the receiver by looking at how well it can decode signals under a variety of conditions representative of real operating environments, specifically the radio channel conditions. Many useful tests can be carried out in fairly simple conditions, where the radio channel is represented just by the addition of a random noise signal, known as additive white Gaussian noise (AWGN). The SS will often have the ability to add this itself, with the advantage that this is usually added at baseband, so the RF cabling is kept simple and the power accuracy is much better. A second class of tests, however, use more sophisticated channel simulations, and these require the addition of a fading simulator. A typical test set-up is shown in Figure 8.27.

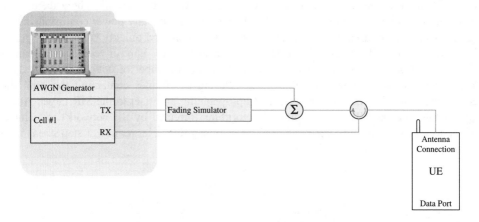

Figure 8.27   System for testing receiver performance

The fading simulator provides simulation of multipath by providing delayed images of the original signal, where the attenuation and the amount of delay are programmable. They can also be Doppler-shifted to simulate movement of the UE. There are a number of multipath profiles standardized by the ITU, which are widely used in for mobile communications testing, and these should be preprogrammed in the fading simulator. In addition, custom profiles can be created or real radio channel profiles recorded and re-created by the simulator. The profiles used in the conformance tests fall into three categories:

1. Multipath, with simulated UE speeds ranging from 3 km h^{-1} simulating a pedestrian user, to 30, 50 and 120 km h^{-1} simulating a vehicle, and to 250 km h^{-1} for a fast train.
2. Moving, which refers to the fact that the reflected path moves in time relative to the direct path.
3. Birth–death, where reflections of the signal suddenly appear and disappear.

For these tests, the receiver performance is tested using the BLER, which is measured using test loop two (from TS 34.109). This loops data back including the original CRC it was transmitted with. The test system can then determine whether a block contained an error by checking it against the CRC. The BLER can also be measured by using RLC-acknowledged mode (AM) and counting retransmissions, but this method cannot be used with the RMC bearers, all of which use transparent mode RLC.

The conformance tests are carried out at different data rates that the UE supports, and at a variety of DPCH E_c/I_{or} levels A typical conformance requirement is expressed in terms of BLER at a given DPCH E_c/I_{or} for a given radio channel profile and RMC bearer, as shown in Table 8.14.

Table 8.14 DCH demodulation parameters and requirements in multipath fading propagation conditions

Data rate (RMC)(kbps)	Propagation conditions	I_{oc}	$\hat{I}_{or}/I_{oc}$	DPCH E_c/I_{or}	BLER
12.2				−15.0 dB	0.01
64				−13.9 dB	0.1
	Pedestrian 3 km h^{-1}, one reflection at 976 ns delay and −10 dB attenuation	−60 dBm/3.84 MHz	9 dB	−10.0 dB	0.01
144				−10.6 dB	0.1
				−6.8 dB	0.01
384				−6.3 dB	0.1
				−2.2 dB	0.01

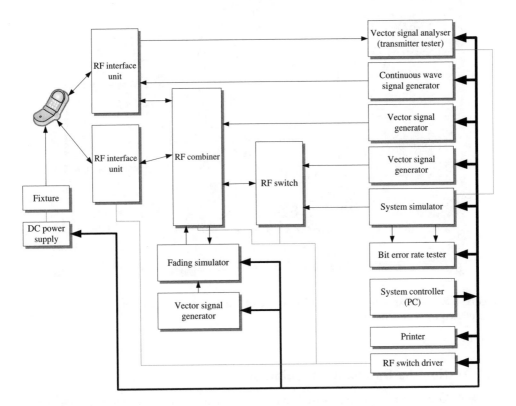

Figure 8.28 Simplified schematic of an RF conformance test system

8.6 RF Conformance Test Systems

An RF conformance test system needs to be able to perform all of the tests detailed in the previous sections. This makes the system quite complex as it has to be able to switch between a number of different configurations, and this switching should be automatic as the user does not want to have to manually change the cabling in between tests. As a number of instruments are needed, together with switching and cabling, the systems tend to be very large. The typical schematic for such a system, showing the variety of instruments needed, is given in Figure 8.28, and an RF conformance test system, the Anritsu ME7873F is shown in Figure 8.29. Due to the cost of these systems, they are not always available to UE developers, and often RF conformance testing is carried out by Test Houses.

Figure 8.29   The Anritsu ME7873F, an example of a fully automated RF conformance test system (reproduced by permission of Anritsu)

8.7 Testing the Baseband in Isolation

So far in this section, we have focussed on testing the radio, as this is the most variable part of the physical layer. The baseband has no specific conformance tests as such, as it is assumed to be tested by implication in the other categories of tests. In general, this is a reasonable assumption. The radio bearer conformance tests in particular will test most of the baseband parameter sets that will be encountered in real networks. However, during development, there are phases, especially early on, where it is useful to be able to specifically test the baseband in isolation. This can be done by connecting test equipment to the modulated I and Q signals. Most usefully, some system simulators will support digital connection to I and Q with the ability to run the simulator at reduced clock speed. This is used during the development of the baseband ASIC. ASIC emulators, which are highly configurable arrays of FPGAs, can be loaded up with the circuit or part of the circuit intended for the ASIC. However these emulators, while being able to execute the functions of the ASIC, are usually unable to run at full speed. The system simulator can be configured directly to send downlink channels and receive uplink channels without the need for upper layer signalling and without layer two. Much of the basic functionality of the transport channel processing, coding and modulation can be tested by sending data directly at the physical layer, looping it back in the baseband ASIC emulation and checking the received data after decoding.

9

Testing of Layer 2

9.1 Introduction

9.1.1 Overview of the RAN Layers

One of the principal requirements on UMTS is that the WCDMA part of the standard needs to be an extension to existing GSM networks rather than require a completely new network. To facilitate this, the network was split into the concept of a core network (CN) and a radio access network (RAN) the CN remaining consistent and the radio access network providing a technology dependent upgrade path. Figure 10.2 shows how the RAN fits into the overall network architecture, and Figure 10.3 shows the structure of the RAN layer 2 and 3 protocols.

9.1.2 Transport Channels and Logical Channels

In UMTS, two types of transport channel exist: common channels and dedicated channels. DCHs are so called because the channel forms a dedicated link to an individual UE and are bidirectional. Common channels are shared by all UEs in the cell and are generally used to carry either broadcast information or the signalling necessary to establish a DCH. However, the common channels are also capable of carrying user traffic. DCHs have the advantage that they offer a much greater degree of control over quality of service, but at the expense of requiring explicit signalling to set up. Common channels are by nature permanently present, or long lived, and do not require explicit set-up, but for traffic that is specific to an individual UE, they have the disadvantage that quality of service is difficult to control as the channel may have to serve more than one UE. The set of standard transport channels are as follows (the channels associated with HSDPA and the E-UL are described separately in Chapter 13):

Testing UMTS: Assuring Conformance and Quality of UMTS User Equipment   Dan Fox
© 2008 John Wiley & Sons, Ltd

Common channels		
BCH	Downlink	Broadcast channel, which is broadcast throughout the cell and is intended for reception by all UEs.
PCH	Downlink	Paging channel, which is also broadcast throughout the cell. This channel can be received by any UE but carries UE specific information. An indication of when information is relevant to a UE is sent separately on the PICH physical channel.
FACH	Downlink	Forward access channel, which mainly carries information intended for specific UEs, discriminated by network-assigned UE identifiers carried in MAC layer headers, but also carries some broadcast information that any UE can receive.
RACH	Uplink	Random access channel, which carries information from specific UEs, identified by network-allocated UE identifiers carried in MAC layer headers. Access to this channel is initiated on demand by UEs and uses a contention resolution protocol.
Dedicated channels		
DCH	Bidirectional	Dedicated channel, which is specific to a UE, and discrimination between different UEs is provided by using a physical channel with a UE specific channelization code in the downlink or scrambling code in the uplink. Codes are allocated by upper layer signalling.

Transport channels carry one or more logical channels, multiplexed together by the MAC layer. In some cases, the mapping is flexible, with the ability to map a logical channel to different transport channels depending on the circumstances. The logical channels can be divided into control channels and traffic channels. The logical channels for standard operation are as follows:

Control channels		
BCCH	Downlink	Broadcast control channel, which carries a set of system information messages which control the behaviour of UEs in the cell and provide information about how to access the other channels. This is mapped to the BCH and under some circumstances can also be used to send system information to UEs in connected mode via the FACH.
PCCH	Downlink	Paging control channel, which carries paging messages for UEs sent by the network to prompt the UE to establish a radio connection. This is carried by its own transport channel, the PCH. The messages contain lists of UE identities being paged, and the UE must parse the list to determine whether it is being paged.

(continued)

CCCH	Bidirectional	Common control channel, which carries RRC layer signalling required to establish a connection, but before the network has been able to assign a connection identity to the UE. In the downlink, this is carried on the FACH and in the uplink, the RACH.
DCCH	Bidirectional	Dedicated control channel, which carries user/network signalling once a connection identity has been assigned. Resolution of data to a specific UE is either implicit, via a dedicated transport channel (i.e. DCH), or in the case of common transport channels (i.e. FACH or RACH), it is explicit in the MAC mapping header.
Traffic channels		
DTCH	Bidirectional	Dedicated traffic channel, which carries user-plane traffic between the network and the UE. This type of channel is only available once a connection identity has been assigned, and addressing of data to each UE is either implicit, via a DCH, or in the case of FACH or RACH, it is explicit in the MAC mapping header.
CTCH	Downlink	This channel is used to carry broadcast data throughout the cell as part of the BMC protocol.

The possible ways in which the MAC can map logical channels to transport channels are shown in Figure 9.1. The choice of multiplexing option is generally set from the RRC, although the RRC can provide the MAC with a range of options and allow the choice to be made based on data availability and radio conditions.

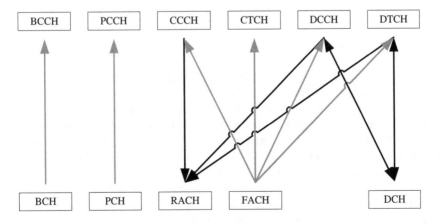

Figure 9.1   Logical to transport channel mappings

9.2 Testing the MAC Layer

9.2.1 Overview of the MAC

The MAC layer has a number of functions, but from the UE perspective, its main role is in the uplink, to multiplex RLC PDUs into transport blocks and deliver them to layer 1 for transport channel processing, and in the downlink, to receive the transport blocks from layer 1 and demultiplex them into RLC PDUs. In the downlink, this becomes a mechanical process of interpreting the TFCI to work out how to perform the demultiplexing, but in the uplink, this ties in some of the key decision making. The MAC considers the physical layer estimation of the available capacity on the radio link, taking into account the network SIR target, and the amount of data in the RLC queues. It decides the TFC to apply, balancing these two requirements. From the network perspective, the MAC adds another key function, in that it provides the multiplexing for different UEs to share the common channels.

The MAC layer is subdivided into a number of entities associated with the different types of transport channel, as can be seen in Figure 9.2. The function of each entity is as follows:

MAC-b	This handles the BCCH/BCH and is located in the Node B. Its functionality is very straightforward and is not usually tested in isolation but will get adequately tested implicitly during tests on the broadcast of RRC system information.
MAC-c/sh/m	This is responsible for mapping to the common transport channels and is located in the RNC. In the case that user traffic maps onto one of the common channels, this entity will be responsible for that mapping functionality and hence has a connection to the MAC-d. The '-sh' and '-m' functionality are related to TDD and MBMS channels, respectively.
MAC-d	This is responsible for managing user-plane data and is also located in the RNC.

9.2.2 MAC Test Methodology

In general, the test system needs to be able to control exactly what is sent at the MAC layer in order to exercise the peer entity in the UE. This would normally require a specialized test implementation of the SS MAC, as it needs to be able to send abnormal PDUs. Fortunately, in UMTS, the flexibility that makes testing such a challenge actually comes to our aid here. Both the RLC and the MAC have the capability to be configured into transparent modes of operation, in that they will not modify PDUs sent to them. This allows the MAC to be emulated at the test case level while allowing the SS to have a standard implementation of layer 2. This is shown in Figure 9.3, where the test script includes a mixture of RRC signalling, to bring the connection up, and MAC peer-to-peer exchanges to perform the test. The RRC signalling uses the normal functionality of the RLC and MAC layers, but the MAC test cases disable any header management within the MAC entity and utilize a transparent

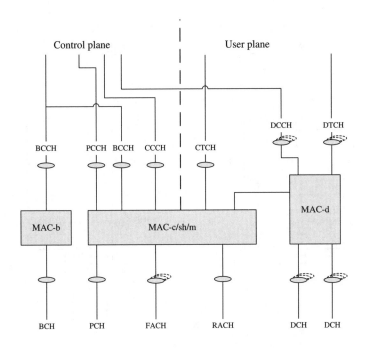

Figure 9.2   Overall MAC architecture (UE side)

mode RLC entity. This allows the PDUs to be built directly in the test script. The UE side consists of a MAC and RLC configured as normal.

9.2.3 The MAC-c/sh/m

In this section we will consider just the MAC-c/sh functionality. The MBMS part ('−m') is outside the scope of this book. This part of the MAC entity is responsible for the following functions relating to the FDD channels on the uplink:

- Multiplexing of the CCCH and any DCHs that are mapped onto common transport channels [target channel type field multiplexing (TCTF Mux)].
- Adding the UE identity for MAC PDUs sent on common channels.
- TF selection.
- Access service class (ASC) selection for RACH access.

On the downlink:

- Demultiplexing of the BCH, CCCH and any DCHs mapped onto the common transport channels (TCTF Demux).
- Discrimination of data for the UE based on the MAC header address field information.

This is shown in Figure 9.4.

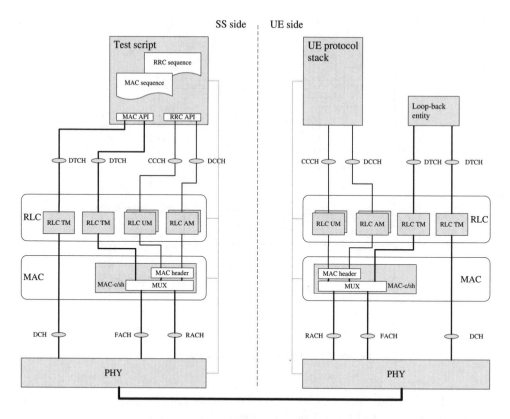

Figure 9.3 Methodology for testing the MAC layer

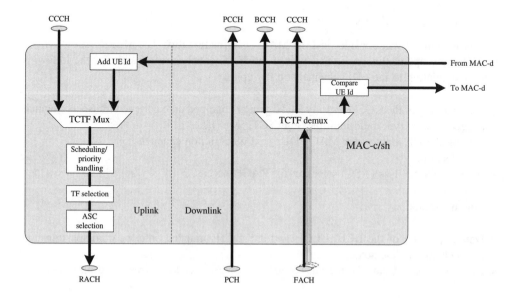

Figure 9.4 Architecture of the MAC-c/sh (UE side)

9.2.4 The MAC-d

The MAC-d is responsible for the management of the dedicated logical channels, so from the UE perspective, the MAC-d entity is active once it is in connected mode. The MAC-d is also located in the RNC but is in the serving RNC, which manages the traffic connections and the associated link security context, so it is possible for the MAC-d to be remote from the MAC-c/sh, which is in the controlling RNC. This part of the MAC entity is responsible for the following functions on the uplink:

- Selection between common and dedicated transport channels, depending on the RRC state.
- Mapping and multiplexing of logical channels onto transport channels. This involves the addition of a C/T Mux field header when more than one logical channel is multiplexed to a transport channel.
- TFC selection.
- Ciphering of the data part of the MAC PDU if the RLC for the logical channel is in transparent mode.

On the downlink:

- Demultiplexing of logical channels from a transport channel where there are more than one logical channel mapped to the transport channel.
- Deciphering the data part of the MAC PDU if the RLC for the logical channel is in transparent mode.

This is shown in Figure 9.5.

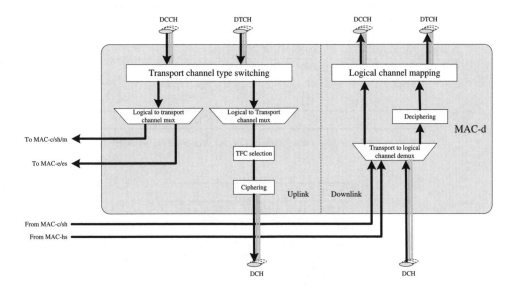

Figure 9.5   Architecture of the MAC-d (UE side)

9.2.5 The MAC Header

In the cases of the RACH and FACH transport channels, a MAC header is needed to provide discrimination of logical channel types and to provide UE discrimination. The first of these is provided by the TCTF, as shown in Figure 9.6, and the coding of this field is shown in Table 9.1.

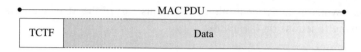

Figure 9.6 General MAC PDU format for RACH and FACH

Table 9.1 Coding for target channel type field of the MAC header

FACH (downlink)				
TCTF			Designation	Field length
b1..b2	b3..b4	b5..b8		
00			BCCH	2
01	00	0000	CCCH	8
		0001	Reserved[1]	8
		...		
		1111		
	01	0000	MCCH	8
		0001	Reserved[1]	
		...		
		1110		
		1111	MSCH	
	10		MTCH	4
	11		Reserved[1]	4
10	00	0000	CTCH	8
	00	0001	Reserved[1]	8
	...	...		
	11	1111		
11			DCCH or DTCH	2

[1] (PDUs with this coding will be discarded by this version of the protocol)

RACH (uplink)		
TCTF	Designation	Field length
b1..b2		
00	CCCH	2
01	DCCH or DTCH	2
10	Reserved	
11	Reserved	

The TCTF field uses a tree-based variable length coding scheme to reduce the protocol overhead for channels that need to carry larger amounts of traffic. To provide future flexibility, there are also a number of reserved fields, and it is important to test that MAC implementations discard PDUs with TCTF values in these reserved spaces.

PDUs carried on the dedicated logical channels (DCCH and DTCH) route through the MAC-d regardless of the transport channel used to carry them. If the radio bearer configuration maps multiple logical channels onto a transport channel, then a header, the C/T field, is used to discriminate logical channels. The output of the MAC-d is shown as case (a) in Figure 9.7 and the coding of the C/T field in Table 9.3. If the PDU is carried on a DCH, then this is the final header, but if it is mapped to a common transport channel, then it is passed to the MAC-c/sh and a further header field is added to provide UE discrimination, as shown in Figure 9.7 case (b). The UE identity is structured as a variable length field of either 18 or 34 bits, depending on the type of address used, which in turn depends on the state of the UE context in the RNC [i.e. whether the UE identity can be resolved down to the cell or the UMTS routing area (URA)]. This is shown in Table 9.4.

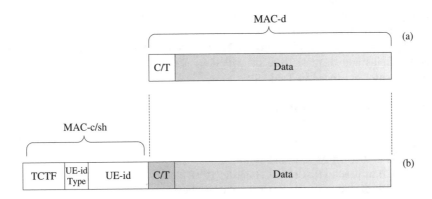

Figure 9.7   MAC PDU format for dedicated logical channels. (a) DCH, (b) RACH/FACH

Table 9.3   C/T field coding for the MAC header

C/T field	Designation
0000	Logical channel 1
0001	Logical channel 2
. . .	. . .
1110	Logical channel 15
1111	Reserved[a]

[a] PDUs containing fields set to values designated as 'reserved' are discarded by current versions of the protocol.

Table 9.4   UE identity field coding for the MAC header

UE-Id type field (2 bits)	UE-Id type	UE-Id field length
00	UTRAN radio network temporary identity (U-RNTI)	32 bits
01	C-RNTI	16 bits
10	Reserved[a]	
11	Reserved[a]	

[a] PDUs containing fields set to values designated as 'reserved' are discarded by current versions of the protocol.

9.2.6 Conformance Tests for the MAC Headers

The normal functioning of the MAC header management is tested implicitly in other signalling tests, as this functionality is quite basic to the operation of the UE and failure tends to have fairly wide effects. The conformance tests are therefore focussed on abnormal conditions. In particular, ensuring predictable behaviour when reserved values are used in the various fields and to make sure the UE does not respond to other UE identities. The relevant test cases are listed in Table 9.5.

Table 9.5   MAC conformance test cases from TS 34.123-1

Test case name	TS 34.123-1 section number
CCCH mapped to RACH/FACH/invalid TCTF	7.1.1.1
DTCH or DCCH mapped to RACH/FACH/invalid TCTF	7.1.1.2
DTCH or DCCH mapped to RACH/FACH/invalid C/T field	7.1.1.3
DTCH or DCCH mapped to RACH/FACH/invalid UE ID type field	7.1.1.4
DTCH or DCCH mapped to RACH/FACH/incorrect UE ID	7.1.1.5
DTCH or DCCH mapped to DCH/invalid C/T field	7.1.1.8

9.2.7 The RACH Procedure

The MAC layer is also responsible for controlling access to the uplink-shared channel, the RACH. The UE needs to be able to gain access to this channel at any time to be able to initiate procedures with the network, but there is a chance that more than one UE will try to access the channel at the same time. In a cellular network it is likely that UEs accessing the same Node B cannot detect each other's signal, so a collision sense protocol, such as collision sense multiple access with collision detection (CSMA/CD) from Ethernet, cannot be used; hence, the MAC uses a contention resolution procedure based on slotted ALOHA, where the terminals transmit blindly on predetermined access slots. In concept, the slotted ALOHA protocol works by providing specific time slots where a terminal can attempt to access the shared media. The receiving entity acknowledges the successful receipt of a message. If two terminals try to access at the same slot, then receiver will not be able to decode either signal and will not acknowledge either terminal. In this case, each terminal will wait for a random number of slots (back-off) and then try again. Based on probability, they will select different back-off times, and their retransmissions will not collide. The weakness of this protocol is that under heavy loading, the number of slots lost through collisions becomes so high that few messages actually make it through. In UMTS, a number of optimizations are used to improve the efficiency under load.

RACH transmissions are split into a preamble part and a message part. The preamble is a short signal transmitted to the network to find out if the channel is clear. The network responds on a special channel, the acquisition indicator channel (AICH). The transmission of the message only proceeds if the network indicates that the terminal may proceed. The preambles do not carry any message information; their purpose is to tell the network that they want access to the PRACH. The preamble is made from a signature, which is one of a set of 16 orthogonal codes. This increases the number of available ALOHA slots by 16, as they are both code and time domain separated. The signature is also transmitted on a different scrambling code to the message part, effectively creating separate channels for the signature and message. Preambles can thus be sent even if another terminal is sending a message. The RACH capacity can be further increased by using additional PRACH channels, and up to 16 of these can be configured for a cell. A further optimization is provided by the use of prioritized ASCs. The network can flexibly map RACH resources (access slots and signatures) to ASCs, such that higher priority ASCs get more resources and therefore more chance of a successful early access.

The basic access protocol is shown in Figure 9.8. The diagram shows two UEs contending for access. The UEs start by randomly selecting an access slot and signature from a group associated with the ASC they have been allocated. In this example, by chance both choose the same slot and signature and a collision occurs. The network does not acknowledge the preambles, so both UEs select a random back-off time and wait. UE #1 times-out first, selects a random slot and signature again and retransmits. In this case there is no collision, and the network acknowledges the preamble on the AICH. The UE can now proceed to transmit the message part. In the meantime, UE #2 has timed-out, selected a slot and signature and retransmitted. The network cannot give it access to the PRACH as it is now allocated to UE #1, so it transmits a negative acknowledge (indicated in this case with a '−' sign). UE #2 will now back-off again to allow for the message transmission and try at a later point.

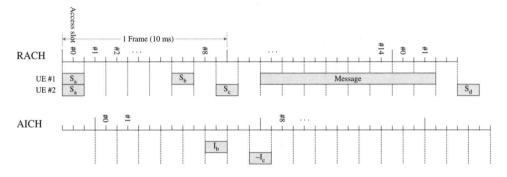

Figure 9.8   Contention resolution with the RACH protocol

This procedure is split between the MAC and the physical layer. The MAC is responsible for selecting the ASC based on the highest priority logical channel (configured by RRC signalling), the PRACH, and for managing the back-off process. The physical layer selects the access slot and signature. It also implements an open-loop power control process, where the power of the first preamble transmission is estimated from a link loss measurement on the pilot channel, and thereafter, preamble retransmissions made when no response is received on the AICH are stepped up in power (Figure 8.15).

9.2.7.1 RACH Testing

Testing the RACH procedure is complicated due to both the random nature of many of the decisions and the subframe timing of key events. The tester has to indicate the slot or access slot number that preambles and messages are received on, as well as the signature chosen. In addition, selection of access slots, signatures and time delays are made on a random basis and require statistical testing to verify fully. The conformance tests take a compromise approach; the test case only checks that the access slot and signature are from the allowed set. The tester also needs to control the AICH transmissions to prevent access indicators from being sent (this can be done by not configuring an AICH) and by forcing a negative acknowledgement (NACK) as a response to a preamble.

9.3 Testing the RLC Layer

9.3.1 Overview of the RLC

The RLC has three modes of operation depending on the level of transmission reliability needed:

TM-RLC	Transparent mode, provided mainly for circuit-switched services, where the RLC entity provides no lost packet detection or recovery, but latency is kept to a minimum and there is no header overhead. In this mode, limited segmentation and reassembly is provided, but in the main framing is provided by the upper layers.
UM-RLC	Unacknowledged mode, providing low latency transport mainly for packet-switched streaming services. Here packet loss detection is provided but not retransmission. This mode also includes segmentation of larger packets and concatenation of smaller ones.
AM-RLC	AM, providing high-reliability transport for packet-switched services where low data loss is more important than latency. This mode provides a level of retransmission of lost data, together with segmentation and concatenation.

The TM and UM-RLC modes have independent paths for the uplink and downlink and can be used unidirectionally, whereas AM-RLC requires a bidirectional channel to handle the retransmission protocol. An RLC entity has a one-to-one relationship with a logical channel, and hence with a radio bearer, and each instance of an RLC entity can be in any of the modes.

9.3.2 TM-RLC

The transparent mode RLC accepts service data units (SDUs) from the source of the radio bearer through TM-SAP, buffers them and on request from the MAC, passes them through a logical channel service access point (SAP). SDUs are expected to be the same size as or exact multiples of one of the configured transparent mode data (TMD) sizes, and one TM entity can be configured with multiple PDU sizes with a resolution down to 1 bit. SDUs which are multiples of PDU sizes are only permitted if segmentation is configured, and then one SDU is sent in each TTI. In the receiver path, PDUs from the MAC are buffered, reassembled if segmentation is enabled and passed up through the TM-SAP. If radio conditions do not allow the immediate transmission of PDUs, they are either discarded, or if timer based discard is configured, they are buffered, the timer is started and transmission is attempted in subsequent TTIs. They are then discarded if they have not yet been sent when the timer expires. The structure of the TM-RLC entity is shown in Figure 9.9.

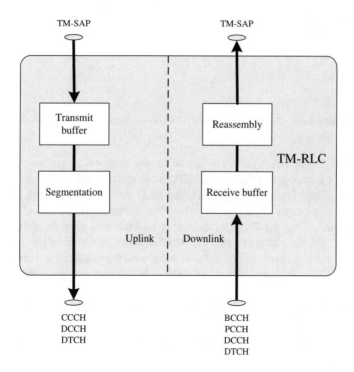

Figure 9.9   Architecture of the transparent mode RLC (UE side)

9.3.3 UM-RLC

The unacknowledged mode RLC, as shown in Figure 9.10, accepts SDUs through the UM-SAP and buffers them. Depending on the transport format selected by the MAC layer,

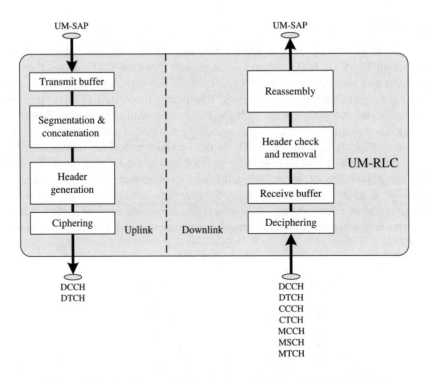

Figure 9.10   Architecture of the unacknowledged mode RLC (UE side)

the RLC will segment SDUs from the queue to the appropriate payload unit (PU) size, or if a PU is not completely filled by an SDU, concatenate further SDUs from the queue until the PU is filled. During this process, the RLC will generate length indicators (LIs) to indicate the ends of any SDUs contained within the payload. The RLC will then add a sequence number, cipher the PU if ciphering is enabled and pass the completed PDU to the MAC entity. On the receiver path, the RLC receives PDUs from the MAC, deciphers the payload and buffers them. In normal operation, missing sequence numbers indicate lost PDUs, and the RLC will reassemble the SDUs that it can and pass them to the upper layers.

The UM-RLC header shown in Figure 9.11 contains the following fields:

- A seven-bit sequence number which provides an index to the PDU and allows the receiving entity to detect lost PDUs.
- An extension or 'E' bit, which when set to '1' indicates that a LI follows, and when set to '0' normally indicates that the data field follows. There is a special case that when the RLC is configured in 'alternative E bit interpretation' mode by the RRC, the E bit following the sequence number is interpreted as meaning that the data field contains a complete SDU without padding. In this mode, a LI is required if the SDU does not completely fill the data field.
- A LI where the first one indicates the number of octets between the end of the RLC header, up to and including the last octet of the SDU, and subsequent LIs indicate the

(a)   (b)

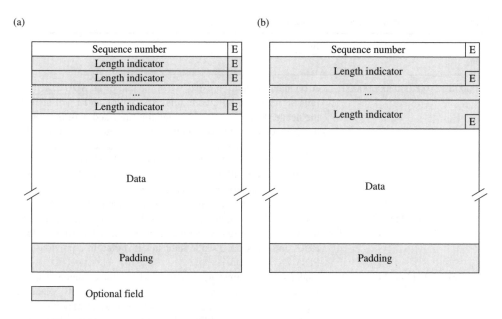

Optional field

Figure 9.11   PDU formats for unacknowledged mode RLC. (a) 7-bit LIs, (b) 15-bit LIs

number of octets between the end of the previous SDU, up to and including the last octet of the SDU to which this LI applies. LIs are normally only included in the RLC PDU in which the SDU ends. That is, if the PDU contains only the beginning part of an SDU it will not have an associated LI, and if a PDU is filled by the mid-section of an SDU, then it will have no LIs. The LI also is followed by an E bit to indicate if another LI follows. The LI can be either 7 bits as shown in case (a), or 15 bits as shown in case (b). The selection of LI size is implied by the largest payload size the RLC is configured for. Seven-bit LIs can only span 125 byte payloads, so larger payloads require 15 bits, even if the SDU terminates within the first 128 bytes.

A number of LI values are reserved for particular boundary situations, and their interpretations are shown in Table 9.6. These are particularly important for testing as they cannot be guaranteed to be exercised during normal operation of the RLC.

9.3.4 AM-RLC

The AM-RLC, as shown in Figure 9.12, is the most complex of the modes and provides a more reliable transport through the use of an automatic repeat request (ARQ) protocol. The transmitter performs segmentation and concatenation of incoming SDUs. Unlike previous modes, the AM only supports one RLC PDU size, and this can only be changed through upper layer signalling. PDUs are sent for transmission and are also placed in a retransmission queue. Under various triggers, the transmitter generates status reports back to the peer RLC, and these are either sent in control PDUs or optionally, if there is enough padding at the end of the data PDU, this padding can be replaced by a status report (in this case, known as a

Table 9.6 LI value interpretations for UM RLC

7-bit value	15-bit value	Interpretation
0000000	000000000000000	A boundary condition exists when the end of an SDU exactly falls at the end of a PDU. In this case, adding an LI will increase the header length, and as PDU lengths are fixed, will then push the SDU end into the next PDU. Of course, this then means the LI cannot be added. This situation is resolved with the use of this LI value. It is placed in the following PDU and indicates that the previous PDU was exactly filled with the last segment of an SDU and has no associated LI in the previous PDU.
	111111111111010 111111111111011	A similar boundary condition exists for 15-bit LIs, which are two octets long, where the SDU finishes in the penultimate octet of the PDU. The first value indicates that the SDU starts in this PDU, and the second value indicates it started in a previous PDU.
1111100	111111111111100	Under some circumstances, particularly on the common channels where messages are sent continually to various targets, when the UE starts to receive PDUs, the first PDU may or may not contain the start of an SDU. The UE has no way of knowing this as it does not have any context from previous PDUs at this stage. However, it is not uncommon for the PDU to contain the start of a message for the UE; for example, if it has just sent an RRC connection request, the network may well be responding. Without some measure, the UE will sometimes miss the first SDU sent to it. This LI, sometimes known as the special LI, indicates that the PDU starts with the beginning of an SDU.
1111101	111111111111101	This indicator is a combination of the previous one and the first boundary condition (above) where the SDU exactly fills the PDU.
1111110	111111111111110	This is used with alternative E bit interpretation and means that the PDU contains a mid-section of an SDU. In some situations, this can be more efficient than the normal E bit use, particularly where the payload size is well matched to the SDU size.
1111111	111111111111111	This indicates that the rest of the RLC PDU is padding. Padding is used to terminate the PDU when there are no further SDUs queued to be sent in that RLC. There is a special case that the addition of this LI pushes the end of the previous SDU to the end of the PDU. In this case the LI is still added, but the padding length is zero.

piggy-backed status PDU). Before transmission, the header field is updated and the PDU is ciphered if it is a data PDU and ciphering is enabled. The receiver gets PDUs from the MAC layer and demultiplexes them into control and data paths. The data PDUs are deciphered and acknowledgements are generated and passed to the transmitter to go to the peer RLC. If the PDU contains a piggy-back status from the peer, this is removed and processed. The PDU is then buffered until all the segments of the SDU are available and it can be reassembled.

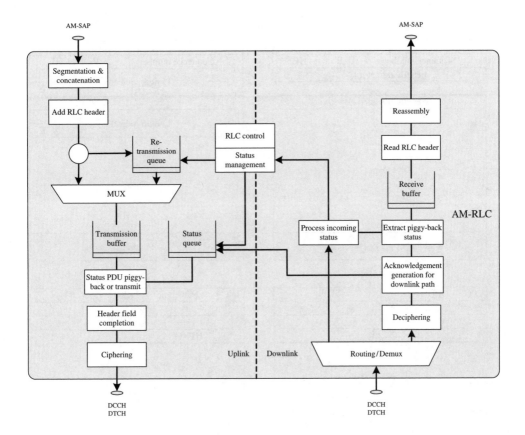

Figure 9.12   Architecture of the acknowledged mode RLC (UE side)

The header for the AM data PDU is shown in Figure 9.13 cases (a) and (b) and contains the following fields:

- A data or control indicator, the D/C field, which indicates whether the PDU contains only control information or not. This field is common between the two PDU types, but subsequently the fields differ. The control PDU format is shown in case (c).
- A 12-bit sequence number, which acts as an index to the PDU and allows the detection of lost PDUs and the resequencing of retransmitted ones.
- A polling request (P) bit, which requests the peer RLC to provide a status report.
- A two-bit header extension (HE) field, which indicates whether the next octet contains data (value '00') or a LI (value '01'). The other two values are reserved and PDUs containing them should be discarded.
- A 7- or 15-bit LI field which has the same purpose and similar encoding to the UM-RLC. The size threshold for automatic selection of 15-bit LIs is when the PDU size is greater than 126 octets (due to the longer header).
- An extension (E) bit indicating that the following octet is either data (value '0') or another LI. There is no alternative interpretation for AM.

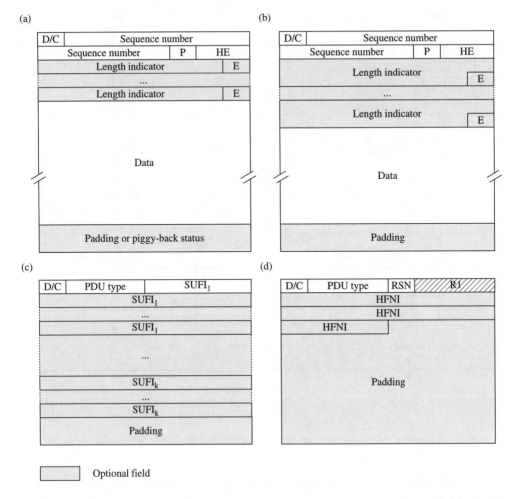

Figure 9.13 PDU formats for acknowledged mode RLC. (a) 7-bit LIs, (b) 15-bit LIs, (c) general control PDU, (d) reset and reset Ack PDU

Again, there are certain boundary conditions that require special LI codes to resolve, as shown in Table 9.7.

To understand the control PDU we first need to review the ARQ protocol.

9.3.4.1 The ARQ Protocol

The basis of the ARQ protocol is quite straightforward. The transmitter keeps a copy of each PDU that it sends. The receiver sends back an acknowledgement for each PDU that it receives. When the transmitter sees the acknowledgement, it removes the copy of the PDU from the queue. If the acknowledgement is not received within a certain time, or an acknowledgement for a subsequently transmitted PDU is received, then the transmitter assumes the PDU was lost and retransmits it. The AM-RLC includes a number of enhancements around this

Table 9.7   LI value interpretations for AM-RLC

7-bit value	15-bit value	Interpretation
0000000	000000000000000	This has the same meaning and covers the same boundary condition as for UM-RLC (Table 9.6)
	111111111111010	Reserved. AM is only used on dedicated logical channels, where the communication is point-to-point. The receiver can always track the start of SDUs
	111111111111011	This has the same meaning and covers the same boundary condition as for UM-RLC (Table 9.6)
1111100	111111111111100	Reserved (see above)
1111101	111111111111101	Reserved (see above)
1111110	111111111111110	This indicates that the rest of the RLC PDU contains piggy-backed status
1111111	111111111111111	This indicates that the rest of the RLC PDU is padding

basic theme to allow the protocol to be optimized towards different qualities of service requirements:

- The transmitter cannot queue PDUs indefinitely, and in fact, PDUs are identified by their sequence numbers. These must be unique at any given time so that the transmitter can identify the correct PDU. This creates a transmission window, where the bottom of the window is set by the oldest PDU transmitted and the top is set by the newest. The window cannot extend more than the size of the sequence number space without duplicating sequence numbers currently in use, so at this point transmission has to stop. In this case, the maximum window size is 4096 (set by the 12-bit sequence number field). The size of this window has a bearing on memory requirements in the UE, as well as on throughput and latency, and is configurable by the upper layers on powers-of-two boundaries.
- The window size of the transmitter can also be dynamically changed by the receiver as a form of flow control. If the transmit window is reduced then the transmitter will not be able to send PDUs beyond the top of the window until any backlog of pending retransmissions is cleared. The window cannot be increased beyond the size configured by the upper layers.
- The transmission of acknowledgements by the receiver takes up some of the link capacity, and therefore it is undesirable to send them too frequently. However, sending them too infrequently increases latency and, as there is a probability that the acknowledgements themselves may get lost, increases the chances of the window stalling. The RLC provides a 'toolkit' approach, where a number of triggers are provided both in the transmitter and receiver that initiate the sending of acknowledgements:
  - Timers: The transmitter can request acknowledgements (status) after a regular time interval, or the receiver can be configured to volunteer status after a regular time interval.
  - PDU counts and SDU counts: The transmitter can request status after a configurable number of PDUs or SDUs have been sent.
  - Window thresholds: The transmitter can request status when the window reaches a certain size.

– The receiver will also send a status report if it receives a PDU with a sequence number more recent than it was expecting.

• The RLC toolbox also has some guard timers that prevent status from being sent or requested too frequently.

9.3.4.2 Error Recovery

If PDUs are held for retransmission for too long, then this can cause a reduction in throughput and increase in latency. This can be avoided by dropping PDUs that cannot get through and relying on the upper layers to recover in the most appropriate way. The upper layers are only interested in SDUs, and any SDUs with segments contained in that PDU cannot be reassembled. Hence it is more efficient to discard all SDUs associated with a dropped PDU. The AM-RLC provides several mechanisms for controlling the discarding of SDUs:

• Timer based, where SDUs are discarded if they have not been successfully transmitted after a certain time. Once the SDU is discarded, a command is sent to the receiver to move the bottom of the receive window up past the PDUs associated with this SDU.
• Discard if the number of unsuccessful retransmissions exceeds a certain value.

The RLC can also be configured to execute a peer reset procedure where both RLCs are brought back to a known state if a PDU retransmission fails a certain number of times.

9.3.4.3 Control Super Fields

The super fields (SUFIs) contain either status reports from the receiver to the transmitter or commands (and their responses) between the peer RLCs. They are defined as follows:

Commands	Window size	This is a command from the receiver to the transmitter and allows the receiver to modify the size of the transmit window to throttle the data flow.
	Move receiving window and move receiving window acknowledgement	This command and response are used to indicate the discard of SDUs.
	No more data	This SUFI indicates that there are no further SUFIs contained in this control PDU.
Status	List and relative list	The receiver needs to report back which PDUs have not been successfully received yet. A variety of formats are available, and the receiver can select the most compact depending on the information it has to communicate. The list and relative list are forms which contain a list of sequence numbers that have not been

(continued)

	correctly received. In the former case this is a straight list of numbers, and in the latter it is the number of the first bad PDU, followed by the distance to the next and so on.
Bitmap	This contains a sequence number relating to the first bit in the bitmap and then a bitmap where each bit represents the next PDU in the sequence. A 1 indicates the PDU has been received and a zero that it has not.
Acknowledgement	This contains a single sequence number and indicates that all PDUs up to but not including that sequence number have been successfully received except for those indicated as errored by any preceding status SUFIs.

9.3.5 Testing the RLC

9.3.5.1 General Considerations for RLC Testing

The methodology for RLC tests is similar in principle to that used for MAC tests. The MAC itself can be left to operate normally in this case, but the SS RLC is placed into transparent mode irrespective of the mode being tested. The test case can then create RLC PDUs exactly as required by the test purpose and transmit them directly to the UE. It can also examine fields directly in received PDUs although this requires an element of decoding the PDU fields. For TTCN-based testing, the SS will provide a set of test suite operations (call-able functions) that will perform this decode. It is particularly necessary for interpreting SUFIs, where the ordering and types used are not predictable. A simplified diagram of this is shown in Figure 9.14.

9.3.5.2 The General RLC Test Procedure

RLC test cases work by establishing a radio access bearer (RAB) with test loop 1 activated. The user-plane bearer (DTCH) uses a UM-RLC or AM-RLC depending on which is being tested, with the SS side configured in transparent mode. The RLC is tested in the user plane, but the assumption is that RLC implementations do not directly take into account whether the connected channels are in the user or control plane. PDUs are sent containing unique data patterns, and when they are returned on the uplink by the loop-back entity, the SS can check the content.

9.3.5.3 Transparent Mode

The transparent mode only provides some limited functionality for segmentation and reassembly. This is not widely used and is not tested by the conformance tests.

9.3.5.4 Acknowledged and Unacknowledged Mode

These are both thoroughly testing using the methods already discussed. The conformance tests include the following:

Unacknowledged mode	Segmentation and reassembly/selection of 7- or 15-bit 'length indicators'
	Segmentation and reassembly/7-bit 'length indicators'/padding
	Segmentation and reassembly/7-bit 'length indicators'/LI = 0
	Reassembly/7-bit 'length indicators'/invalid LI value
	Reassembly/7-bit 'length indicators'/LI value > PDU size
	Reassembly/7-bit 'length indicators'/first data octet LI
	Segmentation and reassembly/15-bit 'length indicators'/padding
	Segmentation and reassembly/15-bit 'length indicators'/LI = 0
	Segmentation and reassembly/15-bit 'length indicators'/one octet short LI
	Reassembly/15-bit 'length indicators'/invalid LI value
	Reassembly/15-bit 'length indicators'/LI value > PDU size
	Reassembly/15-bit 'length indicators'/first data octet LI
Acknowledged mode	Segmentation and reassembly/selection of 7- or 15-bit length indicators
	Segmentation and reassembly/7-bit 'length indicators' / padding or piggy-backed status
	Segmentation and reassembly/7-bit 'length indicators'/LI = 0
	Reassembly/7-bit 'length indicators'/reserved LI value
	Reassembly/7-bit 'length indicators'/LI value > PDU size
	Segmentation and reassembly/15-bit 'length indicators'/padding or piggy-backed status
	Segmentation and reassembly/15-bit 'length indicators'/LI = 0
	Segmentation and reassembly/15-bit 'length indicators'/one octet short LI
	Reassembly/15-bit 'length indicators'/reserved LI value
	Reassembly/15-bit 'length indicators'/LI value > PDU size
	Correct use of sequence numbering
	Control of transmit window
	Control of receive window
	Polling for status/last PDU in transmission queue
	Polling for status/last PDU in retransmission queue
	Polling for status/poll every Poll_PDU PDUs
	Polling for status/poll every Poll_SDU SDUs
	Polling for status/timer-triggered polling (Timer_Poll_Periodic)
	Polling for status/polling on poll_window% of transmission window
	Polling for status/operation of Timer_Poll timer/timer expiry
	Polling for status/operation of Timer_Poll timer/stopping Timer_Poll timer
	Polling for status/operation of Timer_Poll timer/restart of the Timer_Poll timer
	Polling for status/operation of timer Timer_Poll_Prohibit
	Receiver status triggers/detection of missing PDUs
	Receiver status triggers/operation of timer Timer_Status_Periodic
	Receiver status triggers/operation of timer Timer_Status_Prohibit
	Status reporting/abnormal conditions/reception of LIST SUFI with length set to zero
	Timer-based discard, with explicit signalling/expiry of Timer_Discard
	Timer-based discard, with explicit signalling/expiry of Timer_Discard when Timer_STATUS_Prohibit is active

(continued)

Timer-based discard, with explicit signalling/obsolete MRW_ACK
Timer-based discard, with explicit signalling/failure of MRW procedure
SDU discard after MaxDAT-1 number of transmissions
Operation of the RLC reset procedure/UE originated
Operation of the RLC reset procedure/UE terminated
Reconfiguration of RLC parameters by upper layers

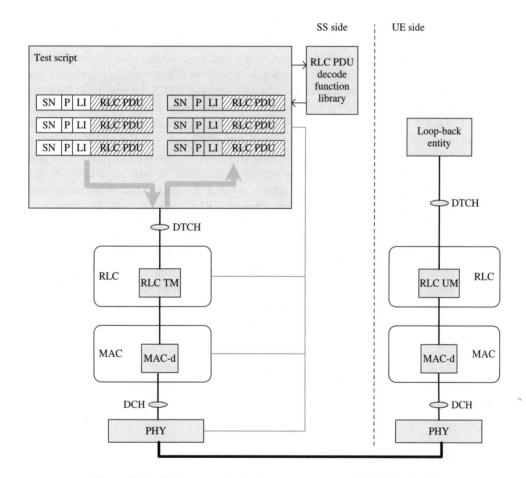

Figure 9.14   RLC test method using a transparent mode RLC in the SS

9.4 Packet Data Convergence Protocol

The PDCP sits only in the user-plane and, as the name suggests, only provides services for
packet-switched traffic. Its two main functions are:

1. To compress Internet protocol (IP) headers. There are two compression methods currently defined although this could be extended in future. The first (Release 99) is the Internet Engineering Task Force (IETF) standard RFC 2507, which is mainly focussed on IP packet headers and TCP headers but can also handle UDP headers. The second (Release 4 onwards) is the IETF standard RFC 3095, also robust header compression (ROHC), which is currently optimized for UDP and RTP, and includes support for the ESP header used with IPSec secure connections although it is being enhanced by the IETF to add TCP support. These compression methods can coexist within a PDCP, and are selected for use through higher layer signalling.

2. To enable relocation of the UE data path from one Serving Radio Network Subsystem (SRNS) to another without loss of data. Under normal conditions, when a UE is handed over from a cell controlled by one RNC to a cell controlled by a different RNC, the new RNC (known as the drift RNC) creates a traffic connection to the original (serving RNC) and the data from the UE is transferred back to it over the I_{ur} connection that links the two. The connection to the CN [the serving GPRS support node (SGSN)] remains unchanged. However, under some circumstances, it is better (or necessary) to move the CN connection to the new RNC and transfer all the context information associated with the UE and its connection between the two. This is SRNS relocation, and while the transfer is occurring the data flow will be interrupted. When used with AM-RLC, the PDCP entity contains support to prevent the loss of packets in this situation. It maintains an internal sequence number for each RLC SDU transmitted. The PDCP entity receives an indication from the RLC once an SDU has been fully acknowledged by the receiving side. During SRNS relocation, the RLCs need to be reset or re-established, and any retransmission state will be lost. The PDCP receivers exchange sequence numbers of the last SDU that was successfully received. This exchange takes place through the RRC signalling and allows the transmitters to discard any that were not yet acknowledged, but were successfully received. If the transmitters and receivers have got out of step for any reason, then they will resynchronize by sending the oldest SDU that has not yet been acknowledged, with a header containing its sequence number.

The PDCP entity can optionally apply a header to each SDU passed to the RLC layer, the format of which is shown in Figure 9.15.

9.4.1 PDCP Testing

The PDCP layer can operate transparently, and the SS can signal a RAB configuration to the UE which applies PDCP, whilst configuring its lower layers to work transparently. This allows the test script to directly apply RLC SDUs with PDCP headers and to receive RLC SDUs from the far end. The data content of the SDUs can also be directly manipulated by the test script. The UE test loop mode 1 is specified such that if the IE 'PDCP Info' is included in the radio bearer configuration then the loop in the UE occurs at the top of the PDCP entity.

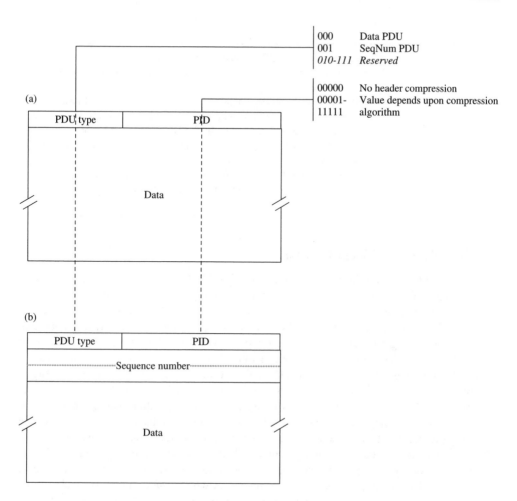

Figure 9.15 PDCP PDU formats. (a) PDCP data PDU, (b) PDCP SeqNum PDU

PDCP conformance testing focuses on the following areas:

Lossless SRNS relocation	The conformance test confirms that when the network sends a PDCP SeqNum PDU, the UE responds with the correct sequence number in a PDCP SeqNum PDU as acknowledgement. More extensive testing could verify that the UE discards PDUs that have not been acknowledged, but are indicated as received in the sequence number exchange, and that the UE will initiate resynchronization if it is sent an invalid sequence number.
RFC2507 compression	Structured testing of compression from the UE is difficult as the compressor has some freedom to decide how to compress, so these tests focus on sending PDUs compressed in various ways

(continued overleaf)

(continued)

| | and ensuring the UE can decode them, loop them back and return them properly compressed. The test just checks that the returned PDUs can be decompressed. |
| ROHC | ROCH testing can be separated into two types: functional testing, where specific IP packets with compressed headers at various states are sent to the UE and the looped back PDU is checked by the tester, and performance testing, where the compressor at the UE side is driven with a sequence of IP headers and the compression ratio achieved is measured. The PDCP specification (TS 25.323) contains IP header sequences and some minimum performance requirements that the UE must meet. |

9.5 Broadcast/Multicast Control

The BMC entity is, from the UE perspective, responsible for receiving messages broadcast using the cell broadcast service. These are generally messages from the network provider sent out over a specific geographic area and intended for all users. The messages are scheduled on a specific channel, the common traffic channel (CTCH), and are the only traffic on that channel. The messages usually have a long duration and are therefore repeated continually on the channel in a round-robin fashion. The schedule of messages currently being broadcast is also sent out periodically on the channel. Once the UE has acquired all the current messages, the BMC entity can implement discontinuous reception (DRX) by only tuning into the channel when the scheduling message is transmitted. The scheduling message contains an indication when any new messages are added, so the UE does not have to continually decode the CTCH to detect when the content is updated. Scheduling takes place at two levels. The CTCH is a logical channel multiplexed onto a FACH. It is only included in the FACH multiplex periodically, and its presence is recognized in the UE through the TCTF field of the MAC header (Section 9.2). The UE can predict in which frames a CTCH block will be present from information provided in the system information blocks (SIBs 5 and 6). On top of the CTCH occasion schedule, there is also a schedule for the broadcast messages.

There is no special test method for BMC. The test system needs to offer support for the CTCH logical channel and for the scheduling of CTCH occasions. It may also offer support for buffering and automatically scheduling of broadcast messages, but the cycle time of these is quite long, and it is possible to manage the schedule from the test script by sending the messages down to the SS repeatedly. Messages are sent using UM-RLC, and the SS RLC needs to support certain levels of control over segmentation and concatenation. For example, SDUs cannot be concatenated and LIs always have to be included in the PDU containing the SDU they relate to.

The BMC tests are difficult to automate; that is, to execute without a test operator present. The broadcast messages are generally text messages similar to SMS messages and usually appear on the display of the UE. The conformance tests focus on sending pre-encoded messages to the UE when it is in the different idle and paged connected states of the RRC, with the test operator checking that the display corresponds to the message content. There is also a test to verify that the UE can correctly pick up changes in the status of messages.

10

Testing of Layer 3

10.1 Overview of the Network Architecture

One of the main design goals for UMTS was that the WCDMA network should seamlessly
extend the existing GSM networks. This led to the network being considered in two parts:
a core network and a radio access network (CN and RAN, respectively). The WCDMA
addition is known as the UTRAN and the legacy GSM RAN as the GERAN. The evolution
of the architecture from GSM to UMTS is shown in Figure 10.1, which is a simplified view

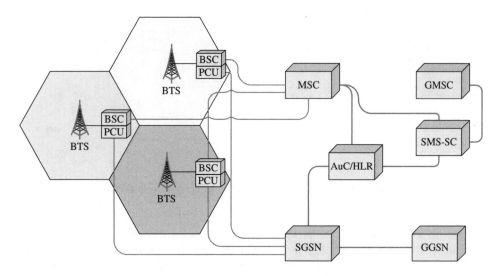

Figure 10.1 Simplified view of GSM network architecture

Testing UMTS: Assuring Conformance and Quality of UMTS User Equipment   Dan Fox
© 2008 John Wiley & Sons, Ltd

of the GSM architecture before the introduction of UMTS, and Figure 10.2, which shows the network with a UTRAN overlay on top of the existing GSM structure.

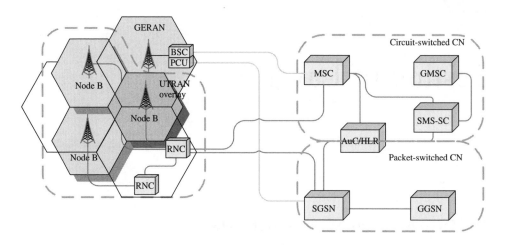

Figure 10.2 Simplified diagram of the UMTS network architecture

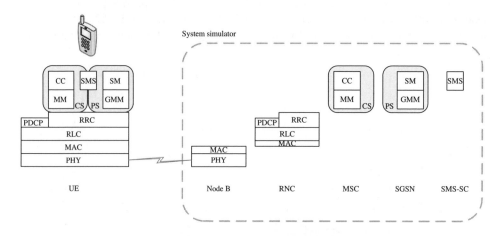

Figure 10.3 UMTS network protocols represented in the SS

From the UE perspective, the protocols are distributed across the various nodes in the RAN and CN. This is shown in Figure 10.3, with the parts of the network protocol of interest to the SS shown in the dashed rounded rectangle. The SS only needs to handle the parts of the protocol that interact directly with the UE. The large amount of signalling that takes place within the network to establish transport bearers, handle security and mobility and so on is transparent to the UE and not represented in the SS. From the perspective of testing, the key network nodes are as follows:

The Node B	Called the base transceiver station (BTS) in GSM, this entity is the controller for an individual cell and is usually located close to the antenna mast. It contains the physical layer processing, including the transport channel coding and multiplexing, the modulation and scrambling and the RF transceiver. Multiple Node Bs connect to a single RNC over the I_{ub} interface. The Node B also contains the MAC-b and from Release 5, implements the MAC-hs for the downlink and the MAC-e/es for the uplink. The latter entities have real-time requirements to process or generate feedback each 2 ms, which makes remote location problematic. The SS will normally have a full implementation of at least one Node B and can usually be extended to add a further two or three full or partial Node Bs, with high-end systems supporting up to six cells in total.
The RNC	The RNC replaces the base station controller (BSC) in GSM but has the capability to control more than one cell. It implements most of the MAC layer, including the MAC-d and MAC-c/sh/m. It also implements the RLC and RRC layers, and in the user plane, the PDCP. The SS will have a standard implementation of the network-side MAC and RLC, and depending on the level of support, it provides for user-plane data, an implementation of the PDCP. The PDCP itself is not used in conformance tests, which concern themselves only with signalling. PDCP testing is done by sourcing and sinking PDCP PDUs from the TTCN test cases. This is also mainly the case for the RRC. A small part of the RRC called the 'thin RRC' (associated mainly with integrity checking and security) is implemented in the SS, otherwise the signalling behaviour is provided from the test scripts.
The MSC	The mobile switching centre is retained from the GSM network. It handles the signalling associated with circuit-switched speech calls and with circuit-switched (CS) data calls (dial-up data services). This entails two signalling entities: MM and call control (CC). In the SS, the CC and MM behaviours are provided from the test scripts. The MSC is a switch, routing speech and data circuits around the network and through a gateway into external fixed line telephony networks. Support for CS traffic is very much dependent on the implementation of the SS. Conformance tests are generally not concerned with simulating user-plane connections, and other than specialized codec testing, at no point is an end-to-end speech call fully tested. However, for most other types of testing (development and IOT, for example), end-to-end connectivity is important. An SS will usually offer some level of support for simulation of a speech path, terminating in an analogue or ISDN handset attached to the tester. It may also offer support for CS data, for example, by connecting the data path to an ISDN circuit.
The SGSN	This network node was introduced into phase 2 GSM networks with the deployment of GPRS, and handles the signalling and traffic for packet-switched data connections. It is responsible for the GPRS mobility management (GMM) and SM protocols, and with the routing of packet data around the network and, through a gateway node, to external packet data networks, in other words, to the Internet. As for the MSC, the SS implements the GMM and SM protocols purely

(continued overleaf)

(continued)

	through test scripts, and again the transfer of end-to-end packet data is not tested during conformance tests. However, this has become increasingly important in other types of testing, as the support for Internet applications and data throughput of UEs are seen as important quality measures.
The SMS-SC	This is responsible for queuing and sending SMS messages. The SMS protocol operates over an MM connection in the CS domain or a PDP context in the packet-switched (PS) domain. In the SS, both are provided from the test scripts, with message text either stored within the test script or with a user interface that allows the message text to be dynamically entered.

Networks are increasingly providing additional application servers with a range of services, such as MMS, IM and PoC. These are treated to varying degrees by SSs, but in general, the trend is towards offering services over a basic IP connection, so testing such features is usually possible with the help of external test applications, or for basic connectivity, even the application servers themselves.

10.2 The RRC

10.2.1 Main Functions

The RRC is logically composed of a number of functional entities. They are introduced here to help put the RRC into context; however, this view of the RRC structure does not have a big impact on testing. The functional entities are as follows:

- A set of routing functional entities (RFEs) that provide routing of higher layer (NAS) messages received on the RRC connection to the relevant NAS protocol entity.
- A broadcast control functional entity (BCFE) responsible for managing state information broadcast to the RRC layer over the network BCH and also for passing the NAS broadcast information to the upper layers.
- A paging and notification functional entity (PNFE), which receives paging messages from the network and notifies the upper layers.
- A dedicated control functional entity (DCFE) responsible for managing the signalling connection for transferring upper layer messages.
- A transfer management entity (TME), which maps traffic from the other entities onto the appropriate RLC SAP.

The RRC is responsible for a wide range of functions related to the operation of the UE within the RAN, meaning that connections with the RRC and mobility within the realm of the RNC are handled at the RRC level. From the UE perspective, the main functions are as follows:

Broadcast information	The network continually broadcasts information relating to the configuration of the network, such as the configuration of the common channels and parameters relating to the operation of connections, reselection criteria for moving to other cells and so on. As this broadcast information does not usually change very dynamically, once this information has been acquired by the UE, it is cached. The network can indicate to UEs when the information changes and which messages have changed.
RRC connection management	The RRC manages the establishment, re-establishment, maintenance and release of RRC connections between the UE and the network based on requests either from the UE upper layers or from the network, for example, via paging. This also includes the management of the mobility aspects of connections.
Radio bearer management	The RRC handles the establishment, reconfiguration and release of radio bearers. Communication between a UE and the network is a two-step process. First, an RRC connection is set up. This allows the transfer of basic signalling information between the two. In the second step, this connection is used to pass signalling to establish a radio bearer, where more radio resources are allocated to the UE, usually including capacity for user-plane data transfer.
Connection mobility	Based on direction or settings from the network, the RRC performs handovers, cell reselections and cell and paging area updates. The RRC also manages measurements relating to these mobility procedures (Section 10.2.8).
Paging/notification	The RRC layer receives paging messages broadcast from the peer entity in the RNC. Paging messages are sent on shared channels, and the RRC must check a list of addresses to identify whether it is being paged.
Routing of NAS signalling PDUs	The RRC receives higher layer signalling using the direct transfer procedure and passes the messages up to the appropriate entity.
UE measurement reporting and control	The UE contains a toolbox of measurements (Section 10.2.8) which can be configured by the network. The RRC gets the measurements from layer 1 and reports them back to the network.
Control of ciphering	The RRC is responsible for enabling and disabling ciphering on the radio link, together with the management of ciphering through various mobility procedures.
Cell selection and reselection	The RRC performs initial cell selection and cell reselection. Some of the parameters to control this are broadcast in the system information, and the RRC will make its decisions based on these parameters and the measurements it takes.
Integrity protection	RRC signalling messages, and by virtue of the fact that upper layer messages are transported in RRC 'wrappers', NAS signalling messages are protected by an authentication code. This prevents unauthorized access to the network.

10.2.2 RRC States

The RRC states (for simplicity, only for UMTS) are shown in Figure 10.4.

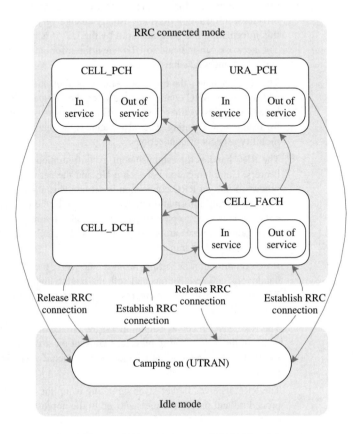

Figure 10.4   RRC states (UTRAN only)

10.2.2.1 Idle Mode

When the UE is powered on it enters idle mode and starts performing a number of tasks that it will periodically execute while it remains in idle mode. These tasks can be broadly grouped as follows:

PLMN selection	If the UE has not selected a PLMN yet, then it will scan to identify a suitable one. It will prioritize the networks it finds, with its home network, if available, as the highest priority and will attempt to camp on to the highest. If necessary, then once the UE is camped on, it will periodically search for a higher priority network. PLMN selection is also performed if the UE looses the signal from the registered network.

(continued)

Cell selection and reselection	Once a PLMN is selected, the UE will search for a suitable cell by making measurements on the air interface. This is a cell on which the UE can at least make speech calls. If it cannot find a suitable cell, then it will camp on to an acceptable cell, where the UE can only make emergency calls (known as limited service). The UE periodically looks for a better cell by measuring the signal strength of the neighbouring cells. The decision on when to reselect to a new cell is guided by parameters broadcast from the network.

Note: When a cell is initially selected or if the UE changes cell to one which is part of a different registration area, the UE will usually perform a location registration. This is a NAS procedure which allows the UE to receive services from the network.

The UE leaves idle mode when a signalling procedure is initiated and attempts to enter RRC connected mode. It can transition to one of two connected states where the connection is active: CELL_FACH or CELL_DCH.

10.2.2.2 CELL_FACH State

In CELL_FACH dedicated radio bearers exist, but these are mapped to the common channels, the FACH for the downlink and the RACH for the uplink. There is no dedicated physical channel. As the bandwidth is limited and is shared by all the UEs in the cell, this state is useful for the transfer of small amounts of data, especially when the transfers are occasional. The connection has an identity allocated locally from the controlling RNC and is unique within the cell. In between transferring data, the UE monitors the broadcast information for changes and measures the strength of surrounding cells. If the signal in the current cell deteriorates too much, the UE can reselect to a better cell and the connection will be transferred, with the UE remaining in CELL_FACH state in the new cell. Data for the UE is broadcast on the S-CCPCH, along with data for all the other UEs in CELL_FACH, so the UE must monitor the allocated FACH channels to look for MAC PDUs addressed to it.

CELL_FACH is reached through the RRC connection setup procedure from idle mode. Once in CELL_FACH, at the RNC's decision (e.g. the connection is idle for long enough), the UE can transition to CELL_PCH or URA_PCH, where there is no active connection. This state can also be reached when the UE moves from CELL_PCH or URA_PCH through the cell update procedure.

10.2.2.3 CELL_DCH State

In CELL_DCH, dedicated radio bearers exist, and they are mapped to a dedicated physical connection – a DCH mapped to a DPCH. The UE makes measurements as instructed by the network and reports them back to the RNC. Cell reselections are not performed, and if the signal in the cell deteriorates, the network may add or remove active radio links (soft handover) or perform a hard handover to another cell.

10.2.2.4 CELL_PCH State and URA_PCH

UMTS recognizes that packet-switched traffic is sometimes bursty in nature, with periods of high activity interspersed with inactivity. Two states, CELL_PCH and URA_PCH, were added to cope with these inactive periods with a reduced level of signalling required to transition back to an active state. In these states, the UE monitors the PCH, and the network will page it if any downlink traffic arrives. The UE also monitors surrounding cells and can reselect if the conditions warrant it. The difference between the two states is that in CELL_PCH, the UE needs to notify the network whenever it changes cell but maintains a current connection identity with the cell (C-RNTI), whereas in URA_PCH, the UE only notifies the network if it moves to a new UTRAN registration area (URA).

10.2.3 Areas and Identities

The UTRAN is divided into a hierarchy of areas, as shown in Figure 10.5. A location area (LA) is made up of a group of URAs and is an area within which the location of the UE is tracked by the CN. A routing area (RA) is the area within which the SGSN tracks the UE and may be the same as or a subdivision of the LA. The URA covers a group of cells where the location of the UE may be tracked during periods of low activity. If the UE is in idle mode, it is not tracked by the UTRAN and has no local identity. Its location, once it has registered, is known only to the MSC for CS and the SGSN for PS and will be a CN identity, such as the international mobile subscriber identity (IMSI) or a temporary mobile subscriber identity (TMSI or P-TMSI) if one has been assigned. Once the UE sets up a connection with the UTRAN, then it is assigned a local identity. Within the URA, it is assigned a UTRAN Radio Network Temporary Identity (U-RNTI), and within the cell, it is assigned a cell RNTI (C-RNTI). If there is a high-speed packet connection (HSDPA or E-DCH), then the UE will also have a separate connection identity for these links.

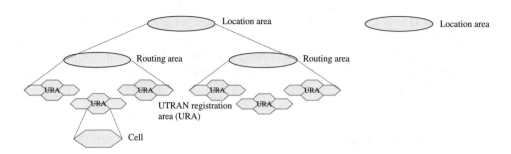

Figure 10.5   Organization of areas in which the CN and UTRAN track the UE

10.2.4 Broadcast of System Information

The UE's RRC is responsible for maintaining an up-to-date set of system information as broadcast on its serving cell. The system information is structured into a number of blocks (SIBs) where each block holds related network information. The SIBs are regularly repeated on the BCH so that a UE can tune in at any time and within one full system information repeat cycle (usually a few seconds) can acquire all the necessary information. To help the UE acquire the most important information quickly, the rate at which individual blocks are repeated can be varied block by block, with the highest priority blocks repeated more frequently within a full cycle. The list of SIBs relating to FDD operation is shown in Table 10.1. Most of the SIBs are only valid within the cell they are broadcast, so the UE must read the system information each time it moves cell. Scheduling of SIBs follows a two-level process. A master information block (MIB) is repeated every eight frames (80 ms), as the UE needs this information to find the rest of the system information. The MIB can directly contain a pointer to other SIBs, giving their position, length and repetition rate. However, the MIB needs to fit within one transport block, so it can be extended by containing pointers to up to two scheduling blocks, which can contain pointers to further SIBs.

Table 10.1 System information blocks for FDD

SIB Type	Description
MIB	Master information block, contains the PLMN identifier and pointers to both SIBs and other scheduling blocks.
Scheduling blocks 1 and 2	Contain pointers to SIBs. A pointer consists of the starting position of the first segment of the SIB, the number of segments and the repetition period.
SIB 1	Contains the LA code, RA code and various parameters relating to the CN. It also contains timer values and key constants mainly relating to upper layer retries or timeouts that apply during idle mode and connected mode procedures.
SIB 2	The URA identity.
SIB 3	Contains the cell identity and the parameters for cell selection and reselection. It also contains an indication as to whether SIB 4 is present.
SIB 4	Contains parameters for cell selection and reselection for connected mode if they are different from idle mode.
SIB 5 (and 5bis)	Defines the configuration of the common channels. A slightly modified version (5bis) is used for bands IV, IX and X. It also contains an indication as to whether SIB 6 is present.
SIB 6	Optionally defines the configuration of the common channels to be used when in connected mode if they are different from idle mode.
SIB 7	Contains parameters controlling PRACH access.
SIB 11 (and 11bis)	Contains parameters for controlling UE measurements, together with definitions of a number of prestored measurements that the UE may be asked to make. In particular, it contains lists of neighbouring cells that need to be monitored by the UE. SIB 11bis is an extension for Release 6 onwards.
SIB 12	If present provides UE measurement control during connected mode.
SIB 16	Contains predefined channel configurations used when handing over from GSM to UTRAN.
SIB 18	Provides PLMN identities matched to the lists of neighbouring cells in SIB 11.

10.2.4.1 Testing Broadcast of System Information

The SS normally caches and automatically broadcasts the system information based on user configurable scheduling information. Encoding the system information messages is a little more complicated than other RRC messages. To support reassembly of segmented messages by the RRC, they are sent in a generic container, which itself is coded as an ASN.1 message. The SIBs are first ASN.1 encoded as a bit string (a sequence of bits), segmented and the segments inserted as bit strings into the container messages, which are then ASN.1 encoded. The SS should provide a set of utilities (test suite operations in TTCN) that take care of the encoding and segmentation.

When it comes to testing SIB reception, for most key parameters, the UE will not behave correctly if the parameters are not interpreted accurately. This will usually surface in tests of the features those parameters apply to. The main area of system information testing is in making sure the UE updates its store when the system information content is modified. The conformance test suite includes paging the UE in connected mode states CELL_PCH and URA_PCH to notify it that the system information has changed. There are also a number of other tests where values in the system information are changed to trigger some action within the UE.

10.2.5 RRC Connection Management

Setting up of an RRC connection is the gateway from Idle Mode to Connected Mode. During Idle Mode the UTRAN does not have any association with the UE; its location is known only to the CN and is stored in the visitors' or home location register (VLR or HLR). The establishment of an RRC connection creates a context for the UE within the RNC, allocates identifiers for the connection and allows the transfer of further signalling between the UE and the network. The RRC connection management procedures are:

Paging	Connections are always set up by a request from the UE towards the network. In the case that the connection is initiated by a network-side event, then the RRC will use a paging procedure. Two procedures are specified: paging type 1, which is used in one of the states where the UE is monitoring the PCH (Idle, CELL_PCH or URA_PCH), and paging type 2, which is used when the UE already has a dedicated connection (CELL_FACH or CELL_DCH). The paging messages contain a cause IE which indicates to the UE why it is being paged. This can be used by the upper layers, for example to decide how to respond or to start up the relevant user application.
RRC connection establishment	This procedure is a three-message exchange, where the UE requests the establishment of a connection by transmitting an RRC CONNECTION REQUEST message on the RACH, the network performs admission control, and if it can accept the connection responds with an RRC CONNECTION SETUP containing the channel configuration information and connection identity on the FACH. The UE then responds with an RRC CONNECTION SETUP COMPLETE on the newly allocated connection (DCCH). If the network cannot accept the connection, it will respond with an RRC CONNECTION REJECT.

(continued)

RRC connection release	This procedure is used to release an RRC connection. It can take place explicitly as a message exchange between the network and the UE or it can be implicit due to a failure of the radio link. If it is explicit, it is always initiated by the network although it may be as a result of the closing of an upper layer link by the UE. The network sends an RRC CONNECTION RELEASE message to the UE, which replies with an RRC CONNECTION RELEASE COMPLETE.
UE capability information	The network needs to know what features the UE supports beyond the minimum set. This information is usually requested by the network early in a signalling procedure through transmitting a UE CAPABILITY ENQUIRY message to the UE, which responds with a UE CAPABILITY INFORMATION message. The network confirms this with a UE CAPABILITY INFORMATION CONFIRM. The procedure can also be spontaneously initiated by the UE if its capabilities change during the life of a connection.
Direct transfer	The direct transfer procedure is used for the exchange of messages from higher layer entities over the RRC connection. NAS procedures start with a request from the UE, even if it is the network that is the initiator, in that the UE will respond to the network paging by establishing a signalling connection. The UE starts the procedure with an INITIAL UPLINK DIRECT TRANSFER, which sets up the signalling connection and carries with it the first NAS message. After that, messages are exchanged from the network using the DOWNLINK DIRECT TRANSFER and from the UE using the UPLINK DIRECT TRANSFER.
Signalling connection release	Signalling connections are usually released implicitly when a NAS procedure reaches a logical conclusion but can also be terminated explicitly with this procedure.
Security mode control	This procedure is initiated by the network and is used to start or reconfigure ciphering and integrity checking. It contains information used to initialize the ciphering and integrity algorithms.

10.2.6 Radio Bearer Control Procedures

10.2.6.1 Radio Bearers

The RRC connection established by the procedures above is used to convey signalling between the network and the UE. It can be pictured as a number of parallel channels, each potentially capable of having a different quality of service (bit rate, BLER and so on). These channels are known as radio bearers and are effectively terminated in the RNC. A radio bearer is identified at the level of a logical channel, meaning that each radio bearer has a unique logical channel and RLC entity associated with it (and PDCP entity if used). The standard recognizes two types of radio bearer: a traffic RB, simply called a radio bearer, and a signalling radio bearer (SRB), five SRBs are defined to carry the different types of network signalling.

- SRB 0: Used for RRC messages sent on the CCCH. These are usually to initiate a connection.
- SRB 1: Used for RRC messages sent using unacknowledged mode RLC.
- SRB 2: Used for RRC messages sent using acknowledged mode RLC.
- SRB 3: Used for carrying NAS messages sent with high priority (equivalent to SAPI 0 on GSM).
- SRB 4: Used for carrying NAS messages sent with low priority (equivalent to SAPI 3 on GSM).

SRB 0 is always present and carries the signalling messages needed to get to an RRC connection. The RRC connection is made up of a combination of the remaining SRBs: SRBs 1, 2, 3 and optionally 4. However, it is limited only to carrying signalling. If a traffic channel is required, then a radio bearer setup procedure is required to create the necessary combination of signalling and traffic radio bearers. The specifications frequently refer to the term 'radio access bearer' (RAB) or RAB combination. This refers to the complete bearer from the CN to the UE, of which the radio bearer is just a part.

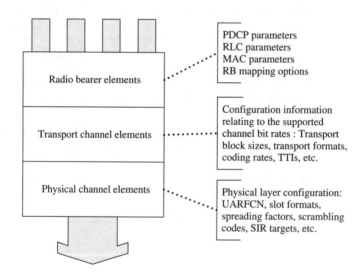

Figure 10.6 Conceptual structure of a radio bearer

Radio bearers are themselves structured into layers, as shown in Figure 10.6, with sets of parameters that define the configuration for each layer. A radio bearer cannot be established in isolation, as it is multiplexed at the MAC with other RBs sharing the transport channel and with other transport channels. Below the MAC layer, the configuration has to take into account the other RBs. Generally, a radio bearer combination will include the four SRBs used in a basic RRC connection, as they provide a path for out-of-band signalling to take place during the call or session. When a RB combination is first set up, the parameters for all the layers are set using the RADIO BEARER SETUP procedure. Once established, a reconfiguration can be targeted only at the layers that require modification, depending on the purpose of the reconfiguration. For example, a handover to another cell can be achieved by reconfiguring only the physical layer parameters, such as the scrambling code.

10.2.6.2 Radio Bearer Management

The radio bearer management procedures are as follows:

Radio bearer establishment	This is performed by sending a RADIO BEARER SETUP message from the network to the UE. The UE responds with a RADIO BEARER SETUP COMPLETE. It defines a complete new configuration of radio bearers, replacing the previous configuration (usually an RRC connection). All the layers have to be defined.
Radio bearer reconfiguration	This is used to reconfigure all three layers of the radio bearers together. It can represent the addition of a new RB, the removal of an old one or a complete change in the types of bearers being used.
Transport channel reconfiguration	This is used to reconfigure the transport channel parameters and can also change the physical layer parameters. It can be used to change the rate or range of supported rates on the radio bearers or the balance of resources between them.
Physical channel reconfiguration	This is used to change the physical parameters without changing the underlying rate of the bearers. It can be used to perform handovers to different frequencies or cells, or for reallocation of the cell's resources. It can also be used to transition the UE between the RRC connected states, particularly where the upper two layers of the radio bearer remain in place, but the physical resources associated with it are released or reacquired.
TFC control	This procedure is used in CELL_DCH or CELL_FACH to flow-control the radio bearers by reducing or increasing the amount of data that can be sent per TTI.
Radio bearer release	This is used to release the resources associated with one or more radio bearers.

10.2.7 Mobility Control

10.2.7.1 Reselection During Connected Mode

When a dedicated connection exists, then movement of the UE between cells is handled by the various handover procedures under direct control of the network. However, if the UE is using the common channels or the connection is inactive, then moving to another cell is the responsibility of the UE and the network needs to be explicitly notified of any change. This is one of the key functions of the cell update and URA update procedures:

CELL_FACH	During CELL_FACH, traffic is flowing on the common transport channels using dedicated logical channels. The flows are identified by a cell-and-UE-unique identifier carried in the MAC header (Section 9.2). When the cell changes, the network may need to relocate the MAC-c/sh to another RNC, it may need to relocate the traffic channel on the I_{ub} interface to another Node B, and it will need to allocate a new identifier to the connection. This is done by sending a CELL UPDATE message to the new cell. The network responds with a CELL UPDATE CONFIRM allocating the new C-RNTI. In addition, if the UE looses the network completely (goes out of service area), then it will start a timer. If it finds the signal again (re-enters the service area) before the timer expires, then it will also perform a cell update, even if it is still in the same cell as before. If the timer expires first, then the UE will return to idle mode.
CELL_PCH	When the UE needs to move the connection to another cell, it will use the RACH procedure (Section 9.2.7) to gain access to the PRACH channel. The UE transitions to CELL_FACH and transmits a CELL UPDATE message to the new cell. The network responds with a CELL UPDATE CONFIRM, which can return the UE to CELL_PCH state. The UE will also perform the same actions on exiting and re-entering the service area as for CELL_FACH.
URA_PCH	The UE will not take any action if the new cell is still within the same UMTS registration area as the previous one. This allows the UE to roam in a wider area without repeated interaction with the network. However, once it moves into a new registration area, it needs to notify the network and acquire a new U-RNTI. This is done by moving to CELL_FACH and sending a URA UPDATE, to which the network responds with a URA UPDATE CONFIRM, assigning the new U-RNTI and if appropriate, returning the UE to URA_PCH state.

In each of these states, the UE can also be configured (through the system information) to periodically perform a cell or URA update even if the cell/URA has not changed.

10.2.7.2 State Transitions from CELL_PCH and URA_PCH

CELL_PCH and URA_PCH states are intended to provide lower power consumption and more efficient use of network capacity for bursty traffic flows, such as Internet browsing. As in idle, the UE can operate DRX, reducing activity much of the time and only powering its receiver briefly to look for paging indicators. However, unlike idle, these states still have active connections and contexts within the network, and the air interface connection can be quickly reactivated. The conditions for transition into these states are network implementation dependent, but the decision to transition to CELL_PCH might, for example, depend on an activity timer and the transition to URA_PCH on the frequency of cell updates. Transition into these states can be achieved using one of the radio bearer management procedures, such as physical channel reconfiguration, but once in the state, there is no direct signalling connection for the transition back out.

Other than the temporary transition to CELL_FACH for cell or URA updating, the transition back to CELL_FACH is usually triggered by the arrival of data bound for the UE at the RNC or by the need for the UE to send data to the network. The first step depends on whether the data is to be transmitted from the network side or the UE side. In the former case, the network will page the UE, causing it to transition into CELL_FACH, and in the latter, the UE will transition itself. The UE sends a CELL UPDATE message on the CCCH over the PRACH. The network responds with a CELL UPDATE CONFIRM, allocating a C-RNTI, either leaving the UE in CELL_FACH or transitioning it to CELL_DCH, and data transfer can restart. The UE responds to the CELL UPDATE CONFIRM depending on the actions instructed by the message. Normally, the identity reallocations will be confirmed with a UTRAN MOBILITY INFORMATION CONFIRM, but if the CELL UPDATE CONFIRM performs a radio bearer, transport channel or physical channel reconfiguration, then the UE sends a confirm message as if one of those procedures were executed directly.

10.2.7.3 Link Failure

The cell update procedure is also used to recover from a failure of the air interface connection when in CELL_DCH. If the physical layer looses synchronization or if the error rate becomes so high that an acknowledged mode RLC gets an unrecoverable error, then the UE will move to CELL_FACH and attempt to perform a cell update.

10.2.7.4 Soft Handover and Active Sets

Soft handover is an important feature in a WCDMA system. UEs that are near the edge of a cell require more power from the Node B and must transmit at higher power themselves. This creates more noise in neighbouring cells, which are operating at the same frequency, and reduces their capacity. The solution to this is to transmit and receive simultaneously from both the serving cell and the neighbour cells. This allows the powers and thus the noise to be kept to a minimum.

At any given time, the UE can therefore have radio links operating with more than one Node B. In fact, in WCDMA, the UE can support up to eight simultaneous connections. Each will carry the same data content, but with different scrambling and channelization codes and with different TPC commands. This set of simultaneous radio links is called the 'active set'. The UE also maintains a list of Node Bs that it measures and reports back the received signal strength to the network. This is known as the 'monitored set'. Based on measurements made on the monitored set, the network decides whether to add radio links into or remove them from the active set. This is done by sending the UE and ACTIVE SET UPDATE message, which contains a list of radio links to add and a list of the ones to remove. The message can both add and remove links at the same time, but the order with which the UE does this is important as it should not get into a situation where, even temporarily, it has no active links or more than eight active links.

From the UE perspective, soft handover only affects the downlink. Macrodiversity, where the network combines the UE signal as received at more than one Node B, to achieve a better SIR performance, can be handled within the network.

10.2.7.5 Hard Handover

Soft handover cannot always be used to handle UE mobility. Cells in the active set need to be on the same frequency, and whether a cell can be added may also depend on network topology and link capacity. There are some circumstances where a hard handover is still required, and with the 'toolbox' approach taken by the RRC layer, hard handover becomes just one case of the more generalized radio bearer management procedures listed earlier in this section.

Generally, a hard handover is achieved by sending the UE a PHYSICAL CHANNEL RECONFIGURATION message, with the physical layer details (scrambling code, channelization code, frequency, etc.) of the new cell. However, using the reconfiguration toolbox, the handover can also be accompanied by a bearer quality of service change or even a complete remapping of the bearers.

10.2.7.6 Inter-RAT Handover

To properly integrate UMTS into a single network from the user's perspective, the ability to hand over between GSM and the UTRAN is needed. Unfortunately, as this needs to integrate with the existing GSM radio signalling, it cannot be handled within the radio bearer management toolbox in the same way as the hard handovers within the UTRAN. Two procedures are defined to facilitate this:

Inter-RAT handover to UTRAN	In this procedure, an INTER SYSTEM TO UTRAN HANDOVER COMMAND is sent on the GSM signalling connection. This message has been added to the GSM radio resource (RR) protocol (TS 44.018) and contains a description of the radio bearers and their associated physical resources in the UTRAN. The UE disconnects from the GSM bearer, retunes to the UTRAN frequency given, locates and synchronizes with the DCH described in the message and sends a HANDOVER TO UTRAN COMPLETE message on the new UTRAN SRB to confirm. The network leaves the old configuration running until it receives the confirmation message on the UTRAN, so that if the handover fails (e.g. the UE cannot find the new cell or get synchronization with the new DCH), then the UE will revert back to the old GSM channel and respond with a HANDOVER FAILURE message.
Inter-RAT handover from UTRAN	In this procedure, when the network judges that a handover to GSM is necessary, it sends a HANDOVER FROM UTRAN COMMAND to the UE. This message is a wrapper for a GSM message, which depends on the circumstances of the handover. For a normal CS speech call, it contains a HANDOVER COMMAND. For a GPRS data session, it contains a PS HANDOVER COMMAND, and for a dual transfer mode call (simultaneous speech and GPRS connections), a DTM HANDOVER COMMAND. The UE disconnects from the UTRAN configuration, tunes to the GSM cell and synchronizes to the indicated channel (frequency, timeslot, etc.). As GSM is a TDMA technology, timing of transmissions is very important. The UE needs to adapt the time at which it transmits its timeslot according to the distance it is from the basestation, otherwise the

(continued)

---

signal could arrive overlapping a timeslot from another UE. This is done by the UE transmitting a series of very short bursts of data, containing a HANDOVER ACCESS message, at the beginning of the nominal timeslot. The network will receive this somewhere later in the timeslot according to how far away the UE is and can estimate the distance according to the delay. The burst is short enough that even delayed, it should not encroach into the next timeslot. The network then responds with a PHYSICAL INFORMATION message containing the initial timing advance for the UE to use. There are also some optimizations of this procedure possible where the network has some knowledge of the distance of the UE from the new basestation. The UE then goes through a layer 2 synchronization process and transmits a HANDOVER COMPLETE message. If the UE fails to get an acceptable signal on the GSM cell, it will revert back to the previous UTRAN configuration and transmit a HANDOVER FROM UTRAN FAILURE message.

There are two further considerations for testing Inter-RAT handovers. First, the signalling from the GSM RR and the UTRAN RRC layers are very different and use different schemes for message definition and encoding. In essence, both the INTER SYSTEM TO UTRAN HANDOVER COMMAND and the HANDOVER FROM UTRAN COMMAND embed a message from the other protocol inside one of their IEs as a binary sequence (bit string). The encoding of this message needs special support within the test system. TTCN provides a generic mechanism for this by allowing different fields of a structure to have different encoding rules, but in practice, this is not widely supported by test systems, and the definition of such a message becomes rather complicated. Instead, this is usually achieved by providing a test suite operation to perform the encoding and decoding, allowing the encoded message to be handled as a bit string within the test case. Second, the UTRAN provides much more flexibility in terms of bearer configuration than GSM. When moving from GSM, a large bearer definition is required, but in fact, the flexibility provided by such a message is of limited value. This has led 3GPP to introduce a more streamlined method using preconfigurations. Here, a number of radio bearer configurations are defined in the core specifications to closely match the GSM bearers. The INTER SYSTEM TO UTRAN HANDOVER COMMAND can simply point to one of these 'well known' preconfigurations rather than contain a full configuration itself. These preconfigurations have to be hard coded in the UE and need careful testing to make sure they are coded correctly.

10.2.8 RRC Measurement Procedures

10.2.8.1 Measurements and Events

To make many of the mobility, state transition and configuration decisions, the network needs information about the radio environment as seen by the UE. This requires the UE to make measurements of its surroundings and communicate these back to the network. In common with many aspects of the standards, a 'toolbox' approach is taken to measurements, with the UE providing a highly programmable set of measurement functions, which are configured

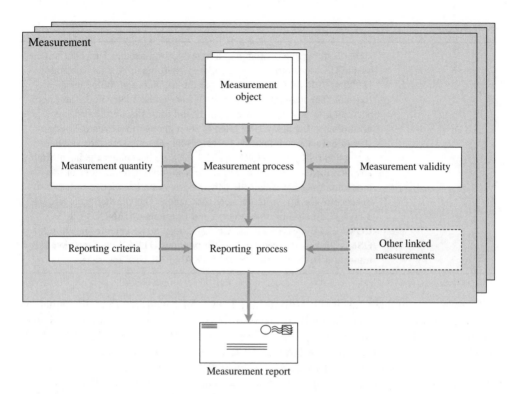

Figure 10.7   Conceptual architecture of the UE measurement handling

by the network to get back the reporting that it needs. The architecture for the measurement toolbox is based around a number of key concepts, as illustrated in Figure 10.7:

Measurement object	This describes what the UE is going to make measurements on. Depending on the type of measurement, this might be a cell, a specific physical channel, an RLC queue and so on. The same measurement can be made on multiple objects.
Measurement quantity	This describes the measurement that will be taken. There are seven basic types or classes of measurement: 1. Intrafrequency 2. Interfrequency 3. Inter-RAT 4. Traffic volume 5. Quality 6. UE internal 7. UE positioning. The quantity measured depends on the type of measurement. For example, for intrafrequency measurements, it may be the E_c/N_0 for the CPICH or its RSCP (the power after despreading).

(continued)

Measurement validity	This is the RRC state or states in which this measurement should be made. This allows the measurements to be set up in one go and left to run. As the UE moves from state to state, the appropriate measurements will be activated and deactivated.
Reporting criteria	This consists of the:

- reporting quantities, which are those measurements to report back and can also include quantities from other linked measurements
- measurement reporting criteria, which are the conditions under which to generate a measurement report. The RRC specification defines a set of events which can be configured as triggers or thresholds to start or stop the generation of reports, and this helps to minimize traffic due to reporting. Reports can also be generated based on timers.

Within the different types of measurement, there are defined quantities that can be measured and events that can be configured. For FDD mode, the principle measurements are detailed in Table 10.2. In addition to these specific measurements, there are some additional

Table 10.2   Measurements and measurement events

Measurement type	Measurement quantity	Measurement object	Related events
Intrafrequency	CPICH RSCP (received signal code power), E_c/N_0 or path loss	Cell	Event 1a: A primary CPICH enters the reporting range (FDD only). Event 1b: A primary CPICH leaves the reporting range (FDD only). Event 1c: A nonactive primary CPICH becomes better than an active primary CPICH (FDD only). Event 1d: Change of best cell (FDD only). Event 1e: A Primary CPICH becomes better than an absolute threshold (FDD only). Event 1f: A primary CPICH becomes worse than an absolute threshold (FDD only).
Interfrequency	CPICH RSCP, E_c/N_0, path loss, or frequency quality (a composite measurement considering all the useable cells on that frequency)	Cell	Event 2a: Change of best frequency. Event 2b: The estimated quality of the currently used frequency is below a certain threshold and the estimated quality of a nonused frequency is above a certain threshold. Event 2c: The estimated quality of a nonused frequency is above a certain threshold.

(continued overleaf)

Table 10.2   (*continued*)

Measurement type	Measurement quantity	Measurement object	Related events
			Event 2d: The estimated quality of the currently used frequency is below a certain threshold.
			Event 2e: The estimated quality of a nonused frequency is below a certain threshold.
			Event 2f: The estimated quality of the currently used frequency is above a certain threshold.
Inter-RAT	GSM Carrier RSSI (received signal strength indication	Cell	Event 3a: The estimated quality of the currently used UTRAN frequency is below a certain threshold and the estimated quality of the other system is above a certain threshold.
			Event 3b: The estimated quality of other system is below a certain threshold.
			Event 3c: The estimated quality of other system is above a certain threshold.
			Event 3d: Change of best cell in other system.
Traffic volume	RLC buffer current (instantaneous) occupancy, average occupancy or variance of buffer occupancy	Transport channel	Event 4a: transport channel traffic volume exceeds an absolute threshold.
			Event 4b: Transport channel traffic volume becomes smaller than an absolute threshold.
Quality	Downlink transport channel BLER	Transport channel	Event 5a: Number of bad CRCs on a certain transport channel exceeds a threshold.
UE internal measurements	UE-transmitted power, UTRA carrier RSSI or UE Rx–Tx time difference	The UE	Event 6a: The UE-transmitted power becomes larger than an absolute threshold.
			Event 6b: The UE-transmitted power becomes less than an absolute threshold.
			Event 6c: The UE-transmitted power reaches its minimum value.
			Event 6d: The UE-transmitted power reaches its maximum value.
			Event 6e: The UE RSSI reaches the UE's dynamic receiver range.
			Event 6f: The UE Rx–Tx time difference for a RL included in the active set becomes larger than an absolute threshold.
			Event 6g: The UE Rx–Tx time difference for a RL included in the active set becomes less than an absolute threshold.

quantities that are also reported back depending on the UE's RRC state, such as timing differences between measured cells.

10.2.8.2 Measurement Control and Reporting

The network has two ways to configure measurements. It can configure a range of standard measurements through the system information (SIB 11). These measurements are then activated in each UE camped on the cell once it transitions into the appropriate RRC state. The network can also configure measurements in a specific UE by sending it a MEASUREMENT CONTROL message. For both methods, reporting is controlled by the reporting criteria, and the way these work depends to some extent on the type of measurement and the type of event used. Figure 10.8 shows a generalized view of the use of a threshold type of reporting criteria. In this figure, an arbitrary quantity is being measured and a reporting criterion is set based on a threshold type of event. At point A, the quantity exceeds the threshold for the first time, and the UE starts a timer. In this case, the quantity does not stay above the threshold for the duration of the time period (B), and no report is generated yet. At point C, the quantity again exceeds the threshold, but this time, it stays above the threshold for long enough, and now at point D, a measurement report is generated. If periodic reporting is configured, then the UE will generate reports as long as the quantity remains in the reporting range each time the period timer expires (E). Once the quantity falls below the threshold (F), then reporting stops.

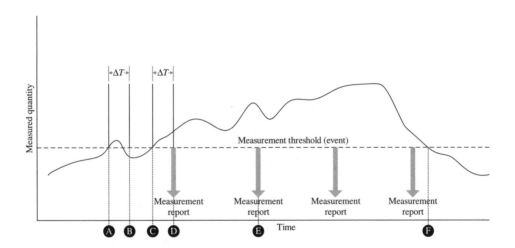

Figure 10.8 Triggering of threshold-based measurement events

10.2.8.3 Testing of Measurements

There are two parts to testing measurements: testing of the signalling associated with measurement control and testing of the accuracy of the measurements themselves. These areas have very different test requirements. The signalling part is best met by the formality of controlled protocol testing, whereas the measurement accuracies require more specialized signal generation and measurement capabilities. In the conformance specifications, these are handled separately; the signalling part is tested as part of the RRC testing and is included in the TTCN ATS for the RRC (part of TS 34.123-3), whereas the measurement testing is covered under the RF conformance tests in section 8 of TS 34.121. These tests form a part of the testing known as RRM and are covered in Section 12.2 in more detail.

10.3 Nonaccess Stratum

The NAS layer is divided into four sublayers. Two of these, the CC sublayer and MM sublayer, are located in the MSC in the network, and the other two, the SM and GMM sublayers, are located in the SGSN. In the network side, these entities are largely independent, but in the UE, there are some linkages, particularly depending on the UE's capabilities. These sublayers do not have any particular architecture for the standards viewpoint, but Figure 10.9 shows a conceptual view with the key subdivisions of functionality.

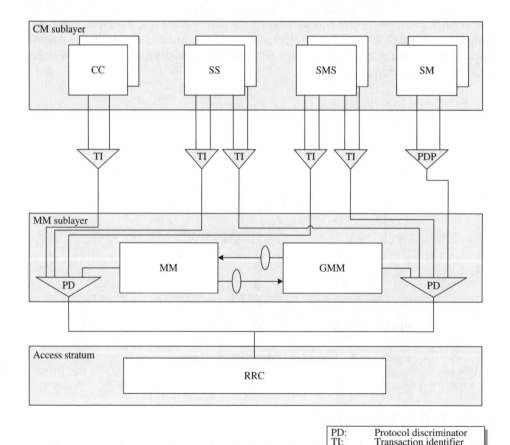

Figure 10.9 Architecture of the NAS

10.3.1 Network Operating Mode

There are two different modes of operation associated with how the network handles paging for UEs capable of both circuit-switched and packet-switched operation:

- Mode I, where the SGSN and the MSC are connected together (via a G_s interface), which allows a measure of coordination between the CS and PS CN domains. The UE can use combined identification, registration and location procedures rather than have to deal separately with each domain controller.
- Mode II, where the SGSN and MSC are not connected, so the UE must signal separately with each node.

10.3.2 Mobility Management

The MM sublayer provides a set of functions that allow the network to track the location of the terminal and create secure connections with it wherever it is located. Its functions can be categorized into those which are specific to the location of the UE within the network and are known as MM-specific procedures, those which are used to provide security and identification support during other types of procedure, known as the MM common procedures. The MM sublayer also provides the connection through which the connection management layer communicates. The MM procedures can be separated into the following five main tasks:

Authentication	This serves two purposes: mutual verification of both the network and the UE, and the allocation of keys for ciphering and integrity. The procedure is initiated by the network with an AUTHENTICATION REQUEST message to the UE, containing a RAND and an authentication parameter (AUTN). The USIM card contains a secret key, known only to the Authentication Centre (AuC) in the user's home network. Using this secret key, the USIM can check that the authentication is genuine. It then uses its secret key and the RAND to generate an expected user response (XRES) back to the network. If this response matches the network's expectation, then the UE (actually the USIM) is genuine and is authorized. The USIM also has two further algorithms that can generate the cipher and integrity keys from the same RAND and AUTN. To reduce the overhead for security, the AuC generates a set of authentication vectors, each consisting of a RAND, AUTN, XRES and the cipher and integrity keys, based on an incrementing sequence number. Subsequent authentications can just take the next vector in the series without having to go back to the AuC.
Location	When the UE is in idle mode, the UTRAN does not store any context information about it. The location of the UE is known only to the resolution of the LA for CS domain and is stored in the VLR. When the UE moves into a new LA, it needs to notify the network, which it does by sending a LOCATION UPDATING REQUEST message with the new LA identity. The network acknowledges this with a LOCATION UPDATING ACCEPT. To prevent stale information accumulating in the network, the UE may also be asked to periodically update its location even if it has not changed (controlled through the system information). This is also done using the LOCATION UPDATING REQUEST message, based on a periodic timer.

(continued overleaf)

(continued)

Identification	To protect the user's confidentiality, most of the signalling between the UE and the network uses a temporary identity, where the relationship between this identity and the UE's real permanent identity, its IMSI, is known only to the network. For circuit-switched services, this identity is known as the TMSI. One reason for this is that ciphering algorithms are easier to break if they contain a known long bit sequence. However, with temporary identities, a third party listening into a network will see only the initial, unsecured, exchange with the IMSI. In a secure network, the TMSI is sent to the UE after ciphering is switched on and is used in all subsequent exchanges. As the TMSI is dynamically allocated and cannot be known outside of the network, it cannot be used to break the cipher algorithm. The TMSI is initially allocated by the network when the UE first camps on and is reallocated when the UE moves into a new LA (in the LOCATION UPDATING ACCEPT). It can also be reallocated when the network deems it necessary (e.g. after a certain time) using the TMSI REALLOCATION COMMAND.
Registration	When the UE first selects the network, an entry for it needs to be created in the VLR and if necessary, its location notified back to its home network. Depending on the network configuration, this can be done explicitly with an 'IMSI attach' or implicitly with the first location update. In both cases, the LOCATION UPDATING REQUEST message is sent.
Connection Management Services	The MM entity is also responsible for managing the signalling connection used by both MM and CC procedures. An MM connection is initiated from the UE by requesting the RRC layer to create an RRC connection and then sending a CM SERVICE REQUEST message. This is either explicitly acknowledged with a CM SERVICE ACCEPT or implicitly acknowledged by the RRC layer switching on ciphering and integrity with an RRC security mode control procedure. A connection can also be initiated by the network by paging the UE with a CN-originated page.

10.3.3 GPRS Mobility Management

The GMM sublayer performs the same set of functions for the PS domain. There are a few minor differences that mainly come about from the ability of this layer to handle combined procedures. In the network, it is optionally possible to link the SGSN and the MSC/VLR together using the G_s interface. In networks with this interface present, the SGSN is able to forward on any relevant update in UE status or location to the VLR, allowing updating of both domains with a single, combined procedure. The key differences between the GMM and MM are as follows:

Authentication	Authentication is performed with the AUTHENTICATION AND CIPHERING REQUEST message, sent from the network to the UE. This message is similar to the CS domain request, with the same authentication parameters, RAND and AUTN. Separate ciphering and integrity keys are used for the PS domain.

(continued)

Location	As PS services can be intermittent in nature, with the UE dropping into 'paged' states when there are significant gaps in the data, the paging traffic load can potentially be much higher than for CS. To prevent paging becoming the bottleneck, LAs can be broken down into multiple RAs. The trade-off is that if the UE is moving, it must update its location more frequently. When the UE moves to a cell in another RA, it sends a ROUTING AREA UPDATE REQUEST message to the network.
Identification	The same concept of temporary identification is used in the packet domain, with the identifier called a packet temporary mobile subscriber identity, or P-TMSI. It is local within a RA and is allocated when the UE registers (sent in the ATTACH ACCEPT) or changes RA (sent in the ROUTING AREA UPDATE ACCEPT). It can also be allocated whenever the network deems necessary, using the P-TMSI REALLOCATION COMMAND.
Registration	The UE first registers on the PS domain by 'attaching' for GPRS services using the ATTACH REQUEST message. In networks operating in mode I, if the UE is not already attached in the CS domain (IMSI attached), then it can perform both attaches in the same ATTACH REQUEST message.
Connection Management Services	The GMM entity creates the GMM connection used by the connection management sublayer for PS signalling. This is done using the SERVICE REQUEST message, which works in an analogous way to the CM SERVICE REQUEST for the CS domain.

10.3.4 Testing MM

MM is a major area for testing, especially within operator acceptance testing. The MM sublayer plays an important part in network security and controlling network access. The conformance tests focus mainly on rejections of the various procedures. There are many reasons the network can reject MM requests, and often these require specific action in the UE, such as marking the PLMN, LA or RA appropriately within the USIM. These tests also often demand multiple cells and sometimes even specific USIM content. Some example MM procedures are given in Section 11.2.

10.3.5 Call Control

The CC sublayer is responsible for the establishment, management and clearing of circuit-switched calls. The protocol is built on the standard ISDN signalling for CC, which helps to map the signalling within the network onto external protocols for calling between networks. CC is based on the state machine shown in Figure 10.10.

10.3.5.1 Mobile-Originated Call

A call is initiated by a setup request primitive from the phone's user interface or similar application and results in the CC entity requesting an MM connection from the MM sublayer,

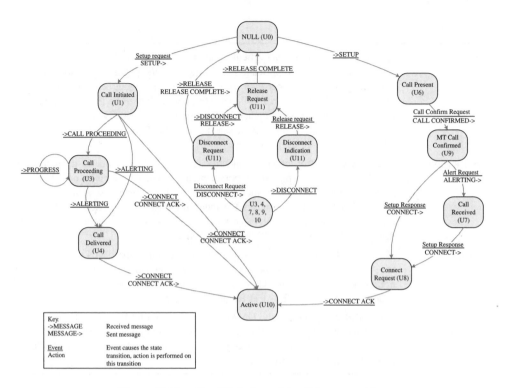

Figure 10.10 Simplified view of the CC state machine

and when it is established, sending a SETUP message and moving to state U1. The network can respond with a CALL PROCEEDING message, which transitions the CC to state U3, or if the call is immediately picked up, a CONNECT message. From state U3, the network may send PROGRESS messages as it routes the call through the network. Once the call reaches the far end, the phone there may ring, in which case an ALERTING message is returned, or it may connect immediately, in which case a CONNECT message is returned. If an ALERTING is received, the CC transitions to state U4, where it will wait for a CONNECT to be returned once the call is picked up. When a CONNECT is received, the CC responds with a CONNECT ACKNOWLEDGE and transitions to state U10, where bearer is connected and the call is active.

10.3.5.2 Mobile-Terminated Call

A mobile-terminated call starts with a paging request from the network which causes the establishment of an RRC connection and then an MM connection, followed by the delivery of a SETUP message from the network. The CC entity transitions to state U6 and indicates to the upper layer that a call is present. The upper layer should request that the call is confirmed; a CALL CONFIRM message is sent to the network and the state transitioned to U9. The phone application will either ring, indicating this to the CC entity with an Alert

Request, resulting in an ALERTING message being sent to the network, or it can immediately answer the call. Either way, once the call is answered, a CONNECT message is sent to the network and the state transitioned to U8, where the CC entity waits for a CONNECT ACKNOWLEDGE from the network. Once this arrives, the bearer is connected, and the CC entity moves to state U10 where the call is active.

10.3.6 Session Management

The SM protocol provides connection management for the packet-switched domain. Packet networks come in two forms: connectionless and connection oriented. The packet network access provided by UMTS is connection oriented, with a dedicated connection provided between the UE at one end and a portal to a service provided by the network operator at the other end. This might be a local service offered to within the network to subscribers, or more often, it is a gateway to the Internet. In terms of the OSI seven-layer model, the session layer sits above the transport layer and is responsible for controlling the connections between hosts, and this is a good reflection of the tasks carried out by the SM protocol. The creation and management of connections uses a connection descriptor called a PDP context. From the UE perspective, the key attributes of the PDP context are as follows:

PDP address	This is the protocol address of the terminating protocol stack within the UE. In the more usual case, where the data protocol is IP, this would be the IP address of the terminating IP stack. A UE can have more than one terminating IP stack. For example, the IP applications embedded within the UE may have one stack, with one IP address, and if it is connected to a PC (for example with USB or Bluetooth), the PC stack will have another IP address. This is covered in more detail below.
NSAPI	This essentially provides the linkage to the bearer in the UE and is an identifier linked to the radio bearer identity carrying the packet data traffic. This identifier is only unique within a UE.
Access point name	This is an identifier for the portal or access point for a service within the operator's network. This is usually a URL, or part of a URL (e.g. without the domain qualifiers), that is resolved by Domain Name Service (DNS) lookup into an IP address. APNs are specific to each operator, and an operator may have more than one APN, each providing access to a different service or set of services. Most commonly, however, the operator will have an APN which relates to the gateway function in the Gateway GPRS Support Node (GGSN) and provides a portal to the Internet.
Quality of service (QoS)	This is a set of parameters that affect how the network handles the user-plane traffic associated with the connection. It allows the applications to specify maximum bit rate, delay class, traffic class, acceptable BERs and packet loss rates, and a variety of related parameters that allow the network to reserve sufficient resources to provide a connection through the network that will meet the needs of the application.

10.3.6.1 Primary and Secondary Contexts

A connection between the UE's protocol stack and the access point always has at least one context which describes the overall connection. This is known as a primary PDP context and is created using the PDP context-activation procedure. If a UE has more than one protocol (e.g. IP) address, each will have a primary context even to the same access point, and each protocol address will have a primary context to each access point it wishes to communicate through. Each primary context can be thought of as describing a pipe through the network. This pipe can itself carry many individual connections to different points beyond the access point. For example, if the access point is the gateway to the Internet, then within this pipe, there can be many TCP or UDP connections to web servers. The primary context also provides an overall QoS for the pipe, and this becomes the default for each individual connection.

It is also possible to differentiate subflows with different levels of QoS. Descriptors for these subflows are called secondary PDP contexts. A number of different ways of differentiating a subflow exist. For example, a subflow can be defined as a connection to a remote IP address or to a set of destination ports. A secondary PDP context contains a set of filters known collectively as a traffic flow template (TFT), which are built up to describe the subflow to which the differentiated QoS should be applied. This in theory allows the connections routing through the GGSN to support delay-tolerant web browsing at the same time as latency tolerant but delay jitter-intolerant streaming applications and so on. Secondary contexts are created using the secondary PDP context-activation procedure. Contexts can also be modified during their active life, if for example the demands of the applications running on the UE change, or for network capacity reasons the bit rate needs to be reduced, or restrictions relaxed. This is done with the PDP context modification procedure. Finally, contexts can be deactivated using the PDP context-deactivation procedure.

10.3.6.2 Testing SM

To perform SM testing a GMM context is needed, and this means that the UE must perform a network attach. The test preamble for creating the GMM context needs to consider variations in UE behaviour due to the network operating mode, the network attach flag and the attach behaviour of the UE, which can often be configured on the device through the user interface. The following areas are tested by the conformance tests:

Normal PDP context activation	This is performed with an attach (by forcing a detach) to also check that if the UE is not attached, it will do this before activating the first context. This is performed for both the network-initiated and UE-initiated cases.
Activation rejections	Two cases of UE rejection of context activation are tested: the UE rejects a context activation from the network if it already has as many active as it can support, and the UE rejects if the network tries to activate a context for an already active connection.
Secondary context activation	Only the UE-initiated secondary context activation is tested. There is also a test for rejection of the activation request by the network.

(continued)

Abnormal cases	The operation of timer T3380: The context-activation procedure is guarded by this timer, which, if there is no response from the network to a PDP CONTEXT ACTIVATION REQUEST, provides timed retries with the eventual abort of the procedure. Collisions where the UE and the network both request activation of primary or secondary contexts, causing the procedures to overlap.
PDP context modification	Both UE and network-initiated PDP context modifications are tested, including rejection by the network for both cases.
Abnormal cases for modification	The operation of the guard timer T3381 for the modification procedure is tested, as is the collision of modification requests from the UE and the network.
PDP context deactivation	There are tests for normal deactivation initiated by both the UE and the network.
Abnormal cases for deactivation	The operation of the guard timer T3390 for the deactivation procedure is tested, as is the collision of deactivation requests from the UE and the network.

10.3.7 Short Message Service

The SMS operates a protocol between the short message centre in the network and the SMS application on the UE. This protocol is divided into two sublayers: the short message relay protocol (SM-RP) and under that the short message control protocol (SM-CP), as shown in Figure 10.11. The protocol sits above the MM and GMM sublayers, meaning that an MM or a GMM connection needs to be established before an SMS transfer can take place.

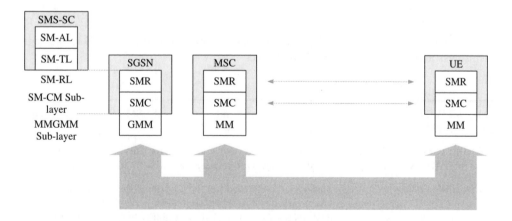

Figure 10.11   SMS protocol entities in the network and the UE

The SMR sublayer is responsible for the transportation of SMS transport layer PDUs (TPDUs). This layer uses services from the short message control (SMC) sublayer, which

handles the acknowledged transfer of SMR PDUs. The various frame formats for the SMC
and SMR sublayers are shown in Figure 10.12. At the SMC sublayer, the CP-DATA PDU is
used to carry SMR PDUs. When the SMC entity is asked to send an SMR PDU, it requests
an MM or a GMM connection, and once the connection is available, it sends a single
CP-DATA PDU and waits for a CP-ACK. Once this is received, it can either send another
SMR PDU or return to idle. The SMR entity relays TPDUs in RP-DATA PDUs (and
optionally in RP-ACK and RP-ERROR PDUs) and also allows the network to determine
whether the UE has available memory capacity to receive an SMS using the RP-SMMA
PDU. If the transmission of a TPDU is requested when the SMR entity is in idle
state, it is packaged in an RP-DATA PDU, given a message sequence number and sent
to the SMC entity for transmission. If the transmission fails, the SMR entity handles
retransmission attempts. When the SMR entity receives TPDUs from its peer, it passes
them to the SMS-TL and then relays acknowledgment or error notifications back from the
SMS-TL.

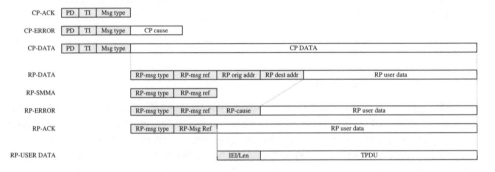

Figure 10.12   SMS PDU types

The short message transport layer is responsible for transporting the short messages
themselves and has the following TPDU types:

SMS-DELIVER and SMS-SUBMIT	These carry the short message from the SMS-SC to the UE and from the UE to the SMS-SC.
SMS-DELIVER-REPORT and SMS-SUBMUT-REPORT	These carry acknowledgement or error causes relating to the transfer of messages between the SMS-SC and the UE.
SMS-COMMAND	These are commands from the UE to the SMS-SC and allow the UE to obtain status information about previously transmitted SMS messages. This includes the ability to delete or cancel them.
SMS-STATUS-REPORT	This is used by the SMS-SC to return the status of previously sent SMS messages. Either as requested when the message is submitted or by a subsequent SMS-COMMAND.

10.3.7.1 Testing of the SMS

The SMS conformance tests are reasonably broad and cover the following areas using both PS and CS deliveries:

- Normal sending and receiving of short messages when the UE is in idle mode.
- Sending and receiving when in a call or an active packet session.
- Reception of a message when memory is full and notification procedures surrounding memory capacity being exceeded.
- Handling of the four special classes of messages (e.g. SMS used to download USIM settings).
- Sending of multiple SMS.
- Overlapped receiving an SMS during the sending of an SMS.

The conformance tests also cover the reception of cell broadcast messages (using BMC), and in particular ensuring that the UE does not continue to indicate the reception of messages it has already received when they are repeated.

Outside of conformance testing, network operators are often keen to test the operation of SMS used to configure features in the UE, as these are often used to automatically enable capabilities that users subscribe to, and UE developers often have extensive interaction tests, where for example, the reception of an SMS during various user interface operations is tested.

11

Testing Protocol

11.1 Protocol Test Systems (System Simulators)

There are a number of protocols that the UE has to deal with; they are mostly complicated and there are many interactions between protocols. This makes protocol testing one of the most complex and time-consuming activities of a UE development. Once the protocol stack is developed and proved, it can be reused in each new platform and model, but maintaining a protocol stack is a continual task. New platforms usually require some adaptations to the stack, which then require development, integration and retesting, but also there is a continual addition of features to the standards, which then necessitate development work in the stack. Protocol test systems are large, complex systems and are a significant investment but will usually last for the lifetime of active protocol development, and this can run for many years. Figure 11.1 shows a typical protocol test system; this model was introduced in 1999 and is still in manufacture, and widespread use, today. To provide such a long lifetime, these systems tend to be modular, often based on an industry standard card format and backplane. This allows parts of the system to be periodically upgraded as the standards evolve.

To provide enough flexibility, a protocol test system has to be able to send any arbitrary sequence of signalling messages in any of the supported protocols. This means the system has to offer some kind of programmatic control to allow the user to create a script. Scripts need to be able to handle the detailed description of messages, down to the content of individual IEs, and they need to allow decisions based on UE actions and the content of messages received from the UE. Ideally scripts should also allow decisions based on user inputs, and they should provide some means for interaction with other test equipment or external programmes or functions. In general, three types of programmatic control are used although individual test systems may not support all of them:

- C and/or C++ language
- TTCN 2
- Graphical.

Testing UMTS: Assuring Conformance and Quality of UMTS User Equipment   Dan Fox
© 2008 John Wiley & Sons, Ltd

Figure 11.1 Commercial System Simulator (Anritsu's protocol test system MX785201, reproduced by permission of Anritsu)

Each has advantages, disadvantages and certain uses. C/C++ language interfaces are popular in some environments due to the widespread familiarity of the language by engineers. Programmatically, the language is powerful and flexible and can be relatively quick to develop scripts with if the environment is well structured, with good libraries and utilities. However, it does have two main disadvantages. While the language itself is widely known, the C environment tends to need a lot of purpose-built utilities and library calls. These essentially form a proprietary 'language' of their own, the semantics and conventions of which then need to be learnt, and because protocol test systems are intrinsically complex, there is a significant learning curve. This can rather defeat the intention of using widely available engineering skills. The second disadvantage is maintainability. The best way to create maintainable test cases is to clearly document the test case before implementing it. Providing the documentation accurately represents the test case, and continues to do so after changes, it is relatively easy for other engineers to become familiar with the code and perform any updates needed. Keeping a close linkage between the documentation and the implementation is the key, and this is easiest to do when there is a good visual association between the two. C programmes do not encourage this; in fact, it is notoriously easy to obscure the logical flow of the programme. As a result, without the very strictest of disciplines, the C-coded test cases soon diverge from the original documentation. Subsequent maintenance work becomes more difficult and often leads to situations, particularly where there is time pressure, when short cuts are taken and alternative versions of functions or data structures written rather than try to pick apart and understand the existing ones. This, in itself, further obstructs future maintenance, and before long the test cases can descend into what programmers refer to as 'spaghetti code'.

This is a long-standing problem and not at all specific to UMTS testing and has led to the development of more standardized 'scripting' approaches to test case development. The most widely used of these is TTCN, which is used across the telecommunications industry. TTCN actually started life as a formal documentation language, but as its structure is well defined, there are a number of commercial tools available that can convert TTCN scripts into executable code. TTCN itself is an abstract language and needs adapting to the test system; however, the tools are well structured to allow this. The test equipment designer provides a set of services and functions that the tool requires from its underlying execution environment. This adaptation and the TTCN tool become part of the protocol test system, and the user can then work at the abstract level without having to understand too much of the specifics of the test system itself. The big advantage of TTCN is that it is easy to relate the documented test case to its TTCN implementation. In fact, this can be done with only a limited knowledge of TTCN. This makes long-term maintenance of test cases much easier and is probably one of the principal reasons for its popularity, particularly with standardization groups, who are faced with the task of maintaining test suites for the long term and in an environment of fairly continual change. Particularly for UMTS, the conformance test specification group selected TTCN version 2, and this has ensured that all the major protocol test solutions offer this capability. The main disadvantage of TTCN is that it is still a specialized language, and finding engineers with the necessary skills is more difficult than with C. The language itself can be learnt relatively quickly by a competent programmer, but it can be difficult to develop a well-structured test suite without at least one reasonably experienced TTCN engineer to make sure the foundations are set correctly.

Particularly in some areas, the problems of finding TTCN-literate engineers, or engineers familiar with the proprietary libraries and interfaces of the C tools, presents a real obstacle. This is especially the case for departments of job functions where testing with protocol testers represents only a part of the normal work load. To service these needs, a couple of more graphical approaches have emerged. One is network emulation, which is discussed in Section 6.3, but has limited flexibility to develop a full range of test scripts. The other is graphical scripting. This is done through libraries of protocol messages and procedures which can be dragged as icons onto a workspace on the screen and then connected together to form complex signalling flows. The advantages of these tools is that they are easy to use and do not require any specialized language knowledge, and as the test sequences are very visual, they are easy to follow and relate back to the test case specification. This by itself simplifies the maintenance task, but the structure of such a tool also lends itself very well to being maintenance friendly. The libraries which underlie the graphical building blocks are provided by the test equipment manufacturer and are maintained centrally, with test cases themselves picking up the updates automatically when they are run with the updated libraries. The main disadvantage of such tools is that they are more costly than the other approaches, as there is significant work involved in developing and maintaining both the libraries of procedures on which they are based and the graphical environment itself. However, they have been steadily gaining in popularity.

11.1.1 Protocols

Part of the complexity of a protocol tester is that it needs to support a range of protocols to test all the parts of the UE protocol stack. For most of these protocols, the tester needs to provide a means to send and receive messages, but for some of the lower layer protocols, the tester needs a complete implementation.In fact, if the protocol itself needs to be tested, then the version in the tester may need to have additional features that allow abnormal states to be tested. Tables 11.1 and 11.2 show the relevant tester protocols and whether the tester needs to have a full implementation.

Table 11.1 Minimum set of UMTS protocols supported by a protocol tester

Protocol	Acronym	3GPP specification	Tester support
Circuit-switched call control	NAS CC	TS 24.008	Message transfer
Circuit-switched mobility management	NAS MM	TS 24.008	Message transfer
Packet-switched session management	NAS SM	TS 24.008	Message transfer
Packet-switched mobility management	NAS GMM or PMM	TS 24.008	Message transfer
Radio resource control	RRC	TS 25.331	Message transfer
Packet data convergence protocol	PDCP	TS 25.323	Full implementation
Radio link control	RLC	TS 25.322	Full implementation
Media access control	MAC	TS 25.321	Full implementation

Table 11.2 Additional protocols to support interworking tests with GSM

Protocol	Acronym	3GPP specification	Type
Circuit-switched radio resource control	NAS RR	GSM 04.18	Message transfer
Packet-switched radio resource management	GRR	GSM 04.60	Message transfer
GPRS subnetwork-dependent convergence protocol	SNDCP	GSM 04.65	Full implementation
GPRS logical link control	LLC	GSM 04.64	Full implementation
Radio link control	RLC	GSM 04.60	Full implementation
Media access control	MAC	GSM 04.60	Full implementation

Figure 11.2 shows the conceptual architecture of a protocol test system. Whilst the real implementation will vary from system to system, this diagram shows the key components required. The parts shown in solid colour constitute the permanent software of the test system. The shaded part is generated from the script written by the customer and will vary according to the tests to be performed. The test platform adapter and the whole environment for running and creating scripts tend to be quite software intensive, but not necessarily real time in that processing deadlines tend to be flexible. Typically, these parts are implemented on a standard Windows computer, either free-standing or using a 19-inch rack mounted PC,

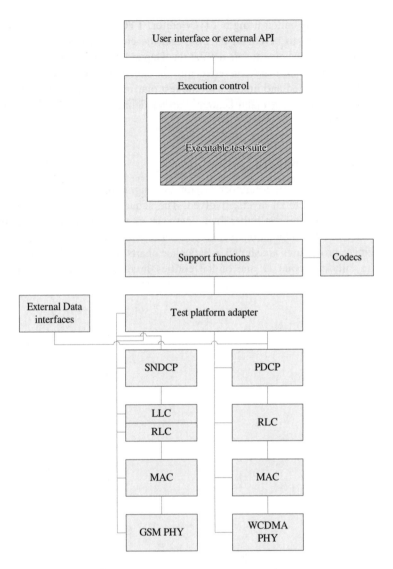

Figure 11.2   General architecture of a protocol tester

for example, integrated into the test system. The layers below the test platform adapter, however, are not so well suited to this type of environment. The MAC layer for UMTS needs to generate transport blocks in the worst case every 10 ms or for HSPA (Releases 5 and 6 onwards) every 2 ms. The MAC also needs some direct interaction with the RLC. Whilst it is possible for these to also reside on the PC, more usually they are implemented in the tester hardware itself, along with the baseband processing and RF circuitry. The choice for the location of the PDCP layer is more open, and can reside on either.

For the tester architecture shown in Figure 11.2, the executable test suite component is generated from a TTCN source file (or set of files), usually through a multistep process.

First, the TTCN source is edited using a TTCN editor. TTCN source files are actually text, but they are not intended to be edited directly, and the code is formatted and tagged in a way that is convenient for software manipulation. Depending on the TTCN tool used, the source code is then either directly converted into an executable or converted into C code. The C code is then compiled and linked with the test platform adapter and support functions supplied by the test equipment manufacturer. Further details of the TTCN 2 language are given in the Appendix.

11.2 Signalling Procedures

In this section, we will review a number of common signalling procedures, looking at the interactions of all the protocol layers involved. The sequence charts are written from the perspective of a UE protocol test rather than a network interaction; the difference being that the sequences include the interactions between the test script and the test system.

In general these procedures are shown as sequence charts (Figure 11.3), where the vertical lines mark an interface point at one of the entities involved in the procedure. Running between the vertical lines arrows are used to represent the transfer of messages between entities.

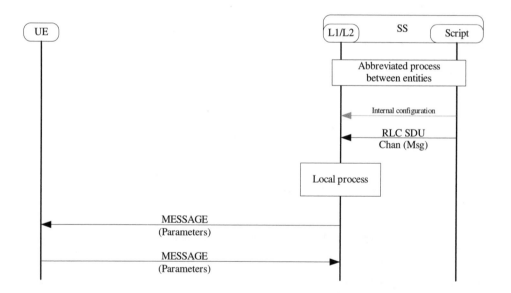

Figure 11.3 Conventions used in the following sequence charts

11.2.1 Setting up Cells

A typical flow for setting up the broadcast of system information from a SS is shown in Figure 11.4. This process can be repeated for each cell required in the test.

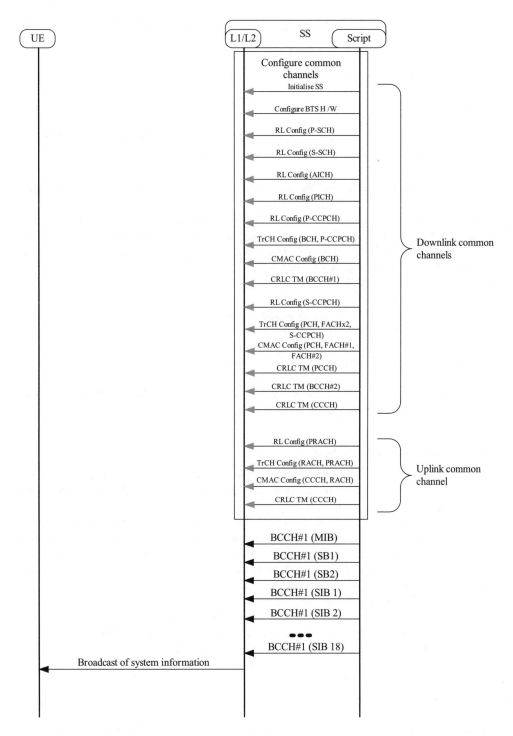

Figure 11.4 Simplified sequence for configuring a cell to broadcast system information

A typical configuration for a cell using a SS would have the following channels on the downlink:

- The primary and secondary synchronization channels P-SCH, S-SCH
- The primary pilot P-CPICH
- The acquisition indicator channel AICH
- The paging indicator channel PICH
- The primary common control channel carrying the broadcast information P-CCPCH
- One or more secondary common control channels S-CCPCHs.

Of these, only the P-CCPCH and the S-CCPCH are connected to higher layers; the others are fixed format and are controlled from layer 1. The P-CCPCH carries a single transport channel, the BCH, on which a BCCH logical channel is mapped. The transport block size and transport formats are fixed for this channel, as is the channelization code and the SF. This is to allow the UE to decode this channel without any prior knowledge. The S-CCPCH configuration is flexible and is signalled to the UE in SIBs 5 and 6. In this example, the S-CCPCH is configured with three transport channels:

1. A PCH carrying the PCCH logical paging channel
2. FACH 1, carrying the BCCH used for broadcast information in connected mode, and the CCCH
3. FACH 2, which will carry traffic from users in CELL_FACH.

On the uplink, a single channel, the PRACH is needed. This carries the RACH transport channel on which the uplink CCCH is mapped.

11.2.2 Configuring Channels

Figure 11.5 shows a sequence chart for configuring a DCH in the SS. The procedure is broadly similar to the flow in the previous section for the common channels, in that first the radio link is configured, followed by the transport channels, the MAC and the RLCs. One key difference is that for the dedicated physical channels, an activation time can be used. This is an SFN on which the configuration is applied. This makes synchronization of procedures much easier. In the example, this is done by finding the current SFN with a query command. The SS will add some time to the returned value to allow the configuration messages to be received by the SS hardware and will then set the activation time in the subsequent configuration messages and the signalling to the UE. This ensures that any change in channel configuration happens simultaneously at the UE and SS, minimizing any loss of data. The configuration generally needs to be done separately for both the uplink and the downlink as the configuration may be unidirectional, with the exception of the AM-RLC, which must be bidirectional.

11.2.3 Location Update

The underlying process for a location update (Figure 11.6) is similar to all the MM-specific procedures: it involves the creation of an RRC connection to bring the UE into CELL_DCH

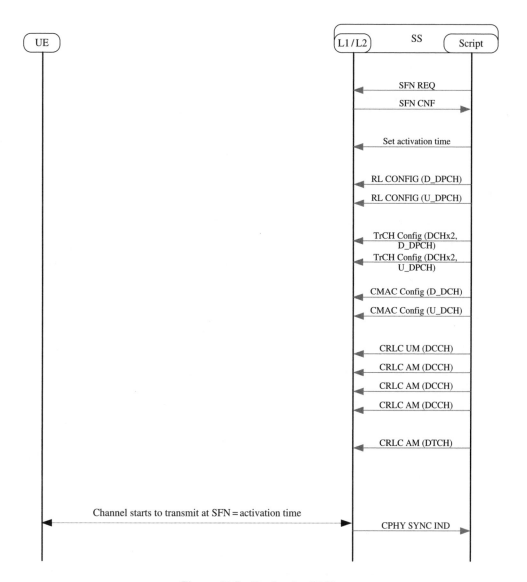

Figure 11.5   Configuring DCHs

state (this example) or CELL FACH state, with a dedicated connection carrying the SRBs. The NAS message, a LOCATION UPDATING REQUEST in this case, is then sent on the high-priority NAS SRB. At this point, the network will usually apply its security procedures, first authenticating the UE and then starting ciphering and integrity protection. All subsequent exchanges then take place on the secured connection. The security settings are applied to the RBs once the SECURITY MODE COMPLETE is received. In fact, the downlink security can be configured immediately following the SECURITY MODE COMMAND, but the uplink needs parameters in the response from the UE. As there are no expected message exchanges between the two, the test case configures both uplink and downlink together.

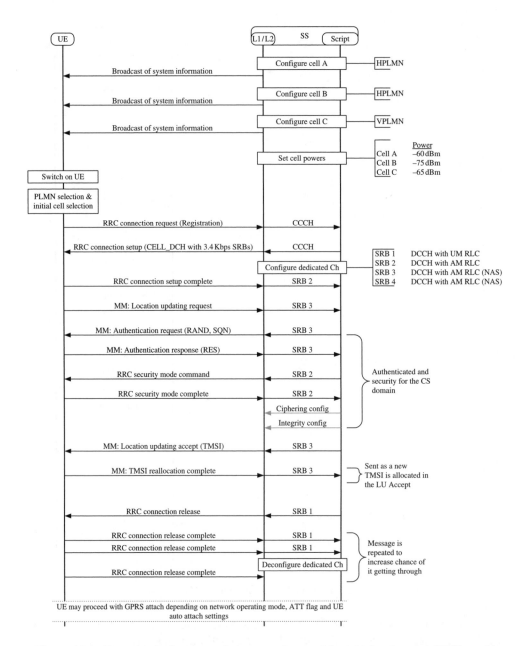

Figure 11.6 Example of a location updating procedure used for initial registration (IMSI attach)

In the case of a location update, the network then responds with a LOCATION UPDATING ACCEPT message. As this example shows the initial UE registration, the network will assign a TMSI in the accept message, and this is confirmed by the UE with a TMSI REALLOCATION COMPLETE. In general, most UEs are capable of both CS and PS operation, and assuming the network is not configured for a combined attach, the UE will now need to perform a GPRS attach. The UE may set the follow-on request (FOR) bit in the initial request, and if the RRC connection is maintained, it will then proceed to perform the GPRS attach. If RRC connection is not maintained, then the UE will set up another RRC connection and perform the GPRS attach. Thus, depending on the scope of the test case, it may need to handle a variety of possible UE behaviours.

11.2.4 Mobile-Originated Circuit-Switched Call

Figure 11.7 shows an example of a CS speech call originated from the UE and disconnected by the UE. This is a single-cell test and the sequence is as follows:

- A single cell is configured and the system information broadcast. The UE is switched on and registers on the cell.
- The test can be automated by sending AT commands to the UE, otherwise the actions must be carried out by a test operator. In this case, an AT command is used to get the UE to dial a number.
- In the UE, the CC entity will request an MM connection, and the MM entity will in turn request an RRC connection. This is initiated by sending an RRC CONNECTION REQUEST using the RACH procedure. The SS responds with an RRC CONNECTION SETUP message, and then configures the downlink accordingly. In this example, the signalling is done in CELL_DCH, using a DPCH with the standard set of SRBs. The RRC CONNECTION SETUP COMPLETE message is sent to the SS once the UE has found synchronization with this DPCH.
- The RRC connection is now established and the MM entity can start to create an MM connection by sending a CM SERVICE REQUEST to the SS.
- The SS authenticates the UE using the test algorithm and enables ciphering and integrity. The security mode command also acts as the acknowledgement of the CM SERVICE REQUEST, completing the establishment of the MM connection. The UE's CC entity can now send the SETUP message.
- The SS responds with a CALL PROCEEDING message which, depending on implementation, may trigger the UE to respond to the ATD command with an 'OK'.
- The SS can now establish the radio access bearer to carry the traffic. In the network, this would allow the reception of progress tones. The RADIO BEARER SETUP (along with the other RB management messages) is in part designed to use a building block approach, in that the parameters in the RB setup allow the specification of the differences from the current configuration applied to that channel. So, in this example, the DCH already has a configuration with four SRBs at 3.4 kbps. These are to be left in situ, and three traffic radio bearers added to carry the three AMR codec subflows carrying the different classes of coded speech bits. The RB set-up message only needs to carry the configuration of the traffic RBs being added, together with those changes at the transport channel and radio link configuration needed to add these into the overall channel multiplex.
- At this point an ALERTING is sent to the UE, followed by a CONNECT. The UE should respond with a CONNECT ACKNOWLEDGE and the audio path is now connected. Depending on the capabilities of the SS, this may involve routing the traffic through a speech codec to an audio port or handset connector.

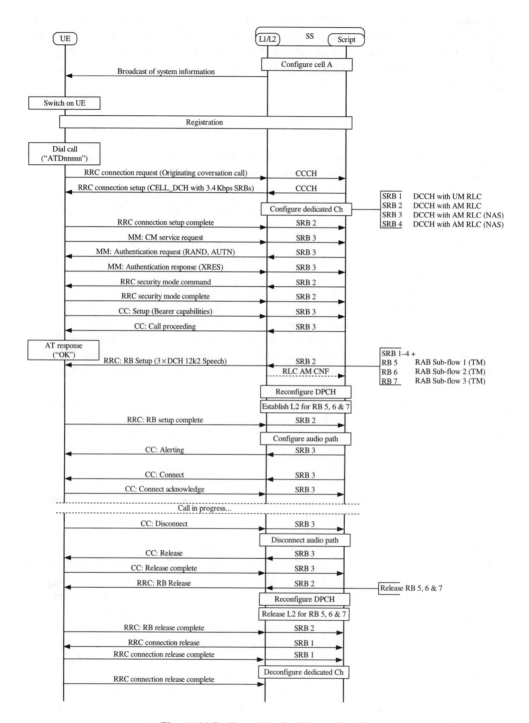

Figure 11.7 Example of a CS speech call

- The call is cleared by one side (in this case the UE) sending a DISCONNECT message. The simulator disconnects the audio path and responds with a RELEASE, which is acknowledged by a RELEASE COMPLETE.
- The SS now releases the resources associated with the speech call by sending a RADIO BEARER RELEASE message. Only the three traffic RBs are released at this stage, leaving the SRBs in place.
- Once the RADIO BEARER RELEASE COMPLETE is received, the RRC connection is closed and the UE returns to idle.

11.2.5 GPRS Attach

Figure 11.8 shows a GPRS attach sent by a UE supporting both CS and PS services, with a network in operating mode II (separate CS and PS procedures) and the ATT flag set (attach required). The choice of when to initiate the GMM attach procedure is down to UE implementation. In this example, the UE is overlapping the two procedures, sending both the CS location update and the PS attach request together. The variation in allowable UE behaviour creates a challenge for test case design as a good generic test should be capable of handling all cases.

The key parts of this sequence are as follows:

- The first part of the sequence is the same as the location update with the exception of the ATTACH REQUEST arriving following the LOCATION UPDATING REQUEST. The test case handles the MM procedure first as this was the first message received.
- Following the TMSI REALLOCATION COMPLETE, the test will continue with the GMM procedure by sending an AUTHENTICATION AND CIPHERING REQUEST with a new authentication vector. The UE responds with an AUTHENTICATION AND CIPHERING RESPONSE and the SS will follow with an RRC SECURITY MODE COMMAND. Once the RRC SECURITY MODE COMPLETE is received back, the SS will configure ciphering and integrity on the SRBs.
- The SRBs now use the ciphering and integrity keys exchanged during the last security mode procedure even though the keys had only just been set, and these keys will be used for any subsequent transfers on the SRBs, irrespective of which domain they are with, until the connection is released or a new security mode procedure occurs.
- The SS responds to the ATTACH REQUEST with an ATTACH ACCEPT, allocating a new P-TMSI, which is acknowledged by the UE with an ATTACH COMPLETE.
- The procedure ends with the release of the RRC connection and the UE returns to idle.

11.2.6 PDP Context Activation

Figure 11.9 shows a PDP context activation procedure, followed by a state change. The sequence is explained as follows:

- The test preamble will perform an IMSI attach and a GPRS attach, and the test itself starts by the UE attempting to establish an RRC connection. In the example, this is triggered by sending an AT command to the UE, but depending on implementation (and sometimes UI settings), context activation can be done automatically.

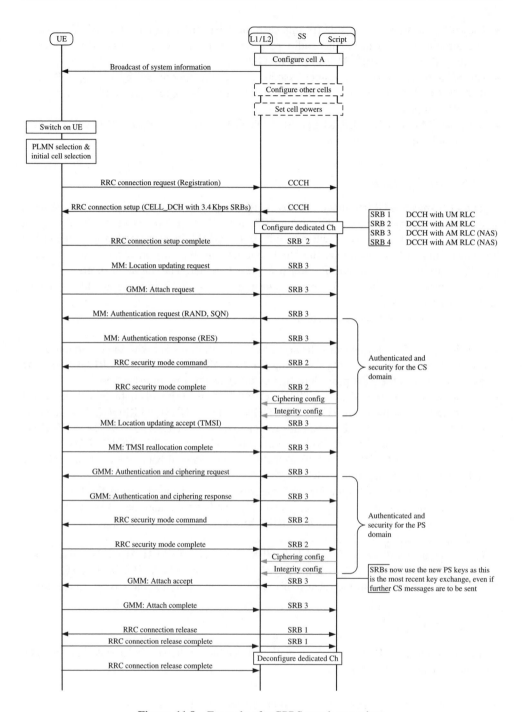

Figure 11.8   Example of a GPRS attach procedure

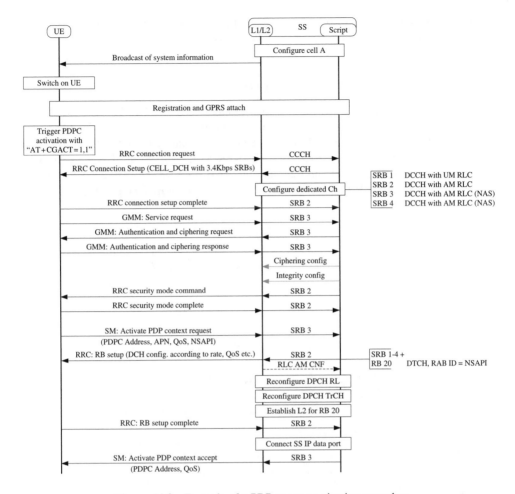

Figure 11.9   Example of a PDP context activation procedure

- Once the RRC connection is established, the GMM entity in the UE attempts to create a GMM connection by sending a SERVICE REQUEST message. This triggers the authentication procedure, and the security mode procedure acts as the confirmation to the UE that the GMM connection is active.
- The UE then sends an ACTIVATE PDP CONTEXT REQUEST with the PDP context address (UE's IP address), the access point name, the QoS that the UE wants and the NSAPI. Usually at this stage, the IP address is omitted and the UE will be dynamically assigned an address by the network. The APN may also be omitted, depending on the network configuration, in which case the network will have a default. The NSAPI is used to relate the PDP context to the associated RAB and becomes the RAB identity.
- The SS will now establish a RAB using a configuration dependent on the requested QoS or the test case purpose, using the radio bearer set-up procedure.
- Once the UE has synchronized to the new radio bearer configuration, it confirms with an ACTIVATE PDP CONTEXT ACCEPT. User data can now flow on the newly assigned

RAB. Depending on the capabilities of the SS, this may include connecting the UE to an IP network or IP application server (e.g. a Web server).

11.2.7 PS Session with State Transitions

This scenario is a continuation of the previous example (Section 11.2.6). In Figure 11.10, we follow the creation of the PS session with a transition to URA_PCH, where the PDP context is still active, but no data is flowing and no physical connection exists. The state transition is effected by sending a RADIO BEARER RELEASE to the UE with the RRC state indicator set to URA_PCH and releasing the RAB used for the PDP context. The UE is bought out of URA_PCH by paging it, in this case due to an incoming speech call.

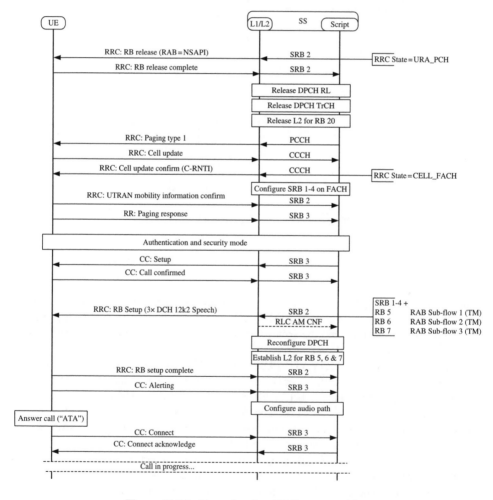

Figure 11.10 Example of an RRC state transition

At this point, the UE has no identity within the cell, so it responds to the page with a CELL UPDATE containing its U-RNTI. The SS sends a CELL UPDATE CONFIRM with a C-RNTI which allows the UE to move into CELL_FACH and use the common channels. The UE confirms this with a UTRAN MOBILITY INFORMATION CONFIRM. From this point, the sequence can continue as usual for a mobile-terminated call.

11.2.8 Soft Handover (Active Set Update)

The scenario in Figure 11.11 shows a speech call with soft handover on two and then three radio links. The test uses three cells in which SIB 11 configures intrafrequency measurements with reporting of event 1a, which occurs when the power of one of the monitored cells enters the reporting range.

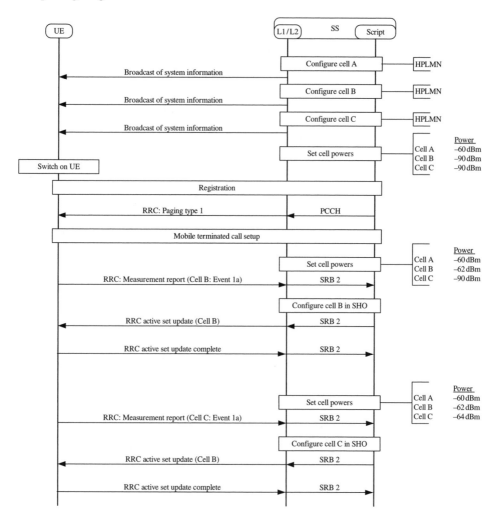

Figure 11.11   Example of an active set update procedure

A speech call is set up on the cell A. Then, the power on cell B is increased to bring it into the reporting range. This triggers the UE to send a measurement report. The SS then configures a downlink DPCH on cell B with the same data content as cell A and adds the new radio link to the active set of the UE. The measurement event and report are not strictly necessary before adding the radio link but ensure that the UE can see the new cell.

12

Testing System Aspects

This section looks at two areas of largely self-contained UE behaviour: idle mode behaviour and RRM. Both of these areas involve the UE making measurements of the radio environment and taking actions based on those measurements. Creating realistic and accurate radio environments for testing purposes requires more complex test systems than for general protocol or RF testing. In particular, these tests often require the test environment to contain many cells. In part because the areas include significant autonomous behaviour in the UE, they are considered very important areas in both conformance testing and operator acceptance testing.

12.1 Idle Mode Procedures

12.1.1 PLMN Selection

Once a UE is switched on, it needs to find a network on which it can obtain service. Even if there are no networks that it can get normal services on, it still needs to try to find a network so that it can make emergency calls. The UE does this by scanning the RF channels in the bands in which it can operate to find the strongest signal. In each RF channel that it identifies a signal, it attempts to find synchronization with the cell and decode the BCH to read the system information and obtain the PLMN identity. Provided the PLMNs meet an acceptable quality threshold, they are put into to a prioritized list for selection. Selection can take place automatically within the UE NAS layer by taking the highest priority in the list and attempting to register, or manually, by presenting the list to the user.

The prioritization of PLMNs in the list is critically important as the operator of the PLMN that is eventually selected will get revenue from use of his network. The prioritization follows strict rules, and the correct application of those rules is subject to close scrutiny in the conformance tests and by the operators. The home PLMN (HPLMN) is, if found, always the top priority in the list. The priority of other PLMNs depends on a variety of factors. The

USIM can contain lists of home operator-preferred PLMNs, for example, where preferential roaming agreements exist or where the networks are part of a common parent company. The user can also configure his own preferences. Where the PLMN is not contained in any of the PLMN selector lists, then it will be added in a random order if the signal quality exceeds a threshold or in signal strength order if it does not.

The UE will work its way down the list attempting to register on the PLMN. If the location update is unsuccessful, the next lower priority is selected. If the UE is unable to select onto its HPLMN, then it will keep trying periodically to reselect to it. The same procedure is also used if the UE loses service on the selected PLMN.

12.1.2 PLMN Types and Cell Types

A number of definitions of PLMN and cell types are used in defining idle mode behaviour. These are summarized as follows:

Home PLMN (HPLMN)	This is the home network as configured in the USIM MCC and MNC fields. This is the highest priority PLMN, and once it is selected, the UE will remain on it unless it loses service (i.e. cannot find any cells) or the user manually reselects to another PLMN.
Visited PLMN (VPLMN)	This is any PLMN other than the UE's HPLMN. If the UE registers on a VPLMN, it will periodically scan for either better PLMNs, for example in terms of its preference lists, or for its HPLMN.
Equivalent PLMN (EPLMN)	A list of EPLMNs is sent by the network in the location update procedure. These PLMNs are treated as having the same priority as the currently selected PLMN for selection and reselection purposes.
Suitable cell	This is a cell which from both a signal quality and a service perspective the UE can camp onto and obtain normal services. It should be a part of the currently registered or selected PLMN or of an equivalent PLMN.
Acceptable cell	This is a cell which has a good enough signal quality but cannot support normal services (e.g. there is no roaming agreement with the home operator). The UE can originate emergency calls on a cell of this type.
Barred or reserved cell	These cells are not in service and are indicated as barred or reserved in their broadcast system information. A normal UE will never select or reselect to them. A reserved cell is reserved for operator use only.

12.1.3 Cell Selection and Reselection

Whenever a PLMN is selected, the process of cell selection starts, as shown in Figure 12.1. The key processes of evaluating cells for selection or reselection depend on a set of rules given in TS 25.304. These rules use parameters which are configurable by the network

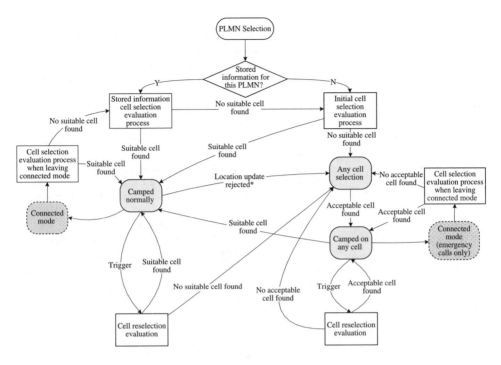

Figure 12.1 Cell selection and reselection process

operator and broadcast in the system information. The key parameters for UTRAN (FDD) are given below. A similar but simpler set of parameters are also used for GERAN cells. Selection and reselection can take place freely between GERAN and UTRAN cells of the same PLMN, with parameters in both RATs that allow the operator to help balance load between the access networks. The key UTRAN parameters are as follows:

$S_{intrasearch}$ and $S_{intersearch}$	Once the UE has selected a cell, it is usually better for it to stay on that cell unless the signal quality deteriorates below a reasonable level. These parameters set thresholds; while the signal quality remains above this level, the UE does not need to search for other cells either on the same frequency or other carriers, respectively.
$Ssearch_{RAT\ n}$	In the same way as the previous parameters, this sets a quality threshold above which the UE does not need to search for other cells in the specified RAT (e.g. GERAN or UTRAN).
$Treselection_s$	This specifies a time interval after which the UE will initiate a reselection evaluation. There are a set of reselection timers for idle mode and the different paging states.

(*continued overleaf*)

(continued)

Qqualmin	This sets a minimum required quality threshold used to determine whether a cell can be considered for selection or reselection. The quality is based on measurements of CPICH E_c/N_0.
Qrxlevmin	This sets a minimum received signal power (in dBm) threshold used to determine whether a cell can be considered for selection or reselection. The received signal power is based on measurements of CPICH RSCP.
Pcompensation	This factor is based on the maximum RACH power allowed in the cell and reduces the chance of a UE trying to camp onto a cell that is too far away for its transmitter to reach. It offsets the selection and reselection ranking of cells where the maximum RACH power is more than the UE can output.
Cell selection and reselection quality measure	This setting tells the UE whether to rank cells for reselection based on their CPICH E_c/N_0 or their CPICH RSCP.
QhystX	As the UE transitions across the boundary between cells, then there is a danger that it will keep reselecting back and forth. To prevent this happening, hysteresis is applied, making the serving cell look more attractive. Different hysteresis measures are applied depending on the reselection quality measure.
QoffsetX	This value applies an offset when ranking the neighbour cells to make them look more or less attractive and can be used to influence the loading on cells.

12.1.4 Hierarchical Cell Selection

UMTS also supports the concept of hierarchical cell structures, where a network of microcells (with small area coverage) is overlaid by an umbrella of larger macrocells. The microcells provide higher capacity and thus higher bit rates for stationary or low-mobility terminals, while faster moving terminals are directed onto the macrocell, where link speeds may be lower, but they will not be continuously handing over between cells.

12.1.5 Testing Idle Mode

Figure 12.2 shows a generic sequence for testing cell selection and PLMN selection. The test script configures three cells for the UE to measure. The cell identities are set according to the purpose of the test case. For example, if the goal is to test HPLMN selection, cell A may belong to the HPLMN and cells B and C to other PLMNs. The power levels on the cells are set such that the UE's decision is deterministic or at least highly probable, taking into account the output level accuracy of the SS, the measurement accuracy allowed to the UE and any cable losses from the actual test set-up. The outcome of the test case (pass or fail) then depends on receiving the RRC CONNECTION SETUP REQUEST on the correct cell. Typically in this type of a test, the result can be influenced by what has happened previously, as the UE will store information to assist in speeding up the cell selection process

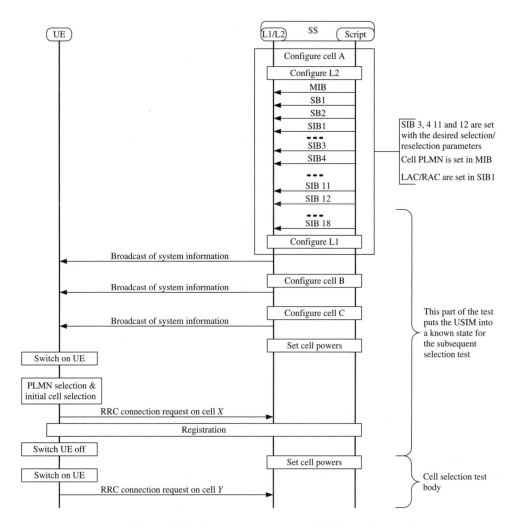

Figure 12.2   Example sequence for PLMN selection

and the USIM will store information about failed registrations. If this is not considered, the tests can be unpredictable in how they run. In the generic flow shown, a preamble is added in the first part of the sequence, which while not directly related to the test itself allows the UE to register successfully and ensures the USIM and other stored information are in a known state. Particular care needs to be taken when testing with barred cells and location update rejections as these can leave the UE in a state where subsequent tests may not run correctly.

Figure 12.3 shows the sequence for a cell reselection test. The preamble is omitted here from the diagram for clarity but may still be required. The first part of the sequence camps the UE onto one cell. The cell powers are changed and the UE should reselect to one of the other cells. The time taken for the reselection may also form a part of the test.

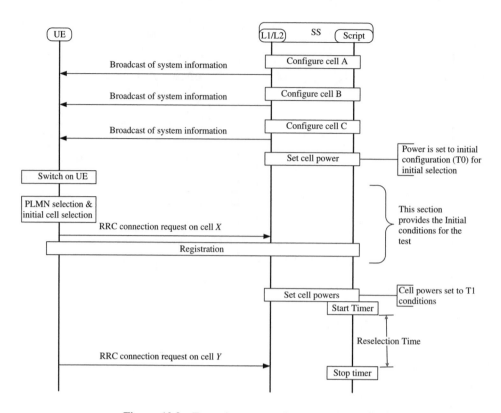

Figure 12.3 Example sequence for a cell reselection

One important consideration in the design of reselection tests is to avoid them inadvertently becoming selection tests. If the power of the serving cell is dropped too low then the UE may assume that it has lost service. While the end result of the test may be the same in that the UE will still attempt to register on the desired cell, the processing and decision points inside the UE may be different.

12.2 Measurements and RRM

The RF conformance test specification, TS 34.121, contains a section on the testing of the RRM support in the UE. As we have seen in Section 10.2.8, the network relies on various measurements taken by the UE. The signalling aspects of many of these areas are handled in TS 34.123, but the measurement aspects themselves are handled in RRM testing. Sometimes, it can be very difficult to separate out the signalling from the measurements, but in general the RRM tests require a more tightly controlled RF environment, and this can also involve more specialized test equipment. A typical RRM test system is shown in Figure 12.4.

In general, where there are areas of overlap, the signalling test scenarios use cells where the PLMN, location/RA, carrier and so on are chosen to test that the application of the rules surrounding the measurement-based decisions made by the UE are in conformance. The RF

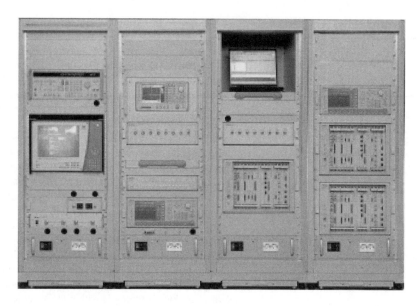

Figure 12.4 An RRM Conformance Test system (Anritsu ME7874F reproduced by permission of Anritsu)

conditions are set to ensure deterministic behaviour, taking into account UE accuracies, SS accuracies and some small safety margin.

The RRM test cases cover the following areas:

Cell reselection, Intersystem (UTRAN to GERAN) reselection	Tests the timing for cell reselection to take place once the RF conditions are set to trigger the reselection. The cell powers are closer to the decision points, and OCNS is used to provide simulation of other users on the cell and to keep I_{or} constant.
Soft handover	Concerned with the timing for the addition of a new radio link, measuring the BLER to determine if the added radio link is operating correctly within the allowed time.
Hard handover and Inter system handover	Tests that the interruption time or handover delay is within the specified limit.
Cell reselections in connected mode	As for the idle mode reselections, these tests check the time for reselection to take place once the RF conditions are set.
RRC re-establishment delay	This checks that the time for the UE to attempt to re-establish an RRC connection following a radio link failure is within the specified limit.
Random access procedure	The RACH procedure is not tested in by the signalling conformance tests. The RRM tests are fairly extensive, checking the power ramp on successive preambles, and the preamble-to-message power difference, the back-off

(*continued overleaf*)

(continued)

	timer when a preamble is responded to with a NACK, and the behaviour when the preamble power reaches the maximum allowed.
TFC selection	On the transmit side, the UE has to balance the amount of data it needs to send in a TTI against the channel quality and its estimate of how much power is required to transmit the TFC. If the required power is more than the maximum the UE is allowed to use at that point in time, then the UE must stop using the TFC. A period of time is allowed for this to happen, as the radio bearer may be carrying streaming traffic (e.g. speech) and the data source may need to reduce its rate. This test checks that TFCs that require excess power are blocked within the specified time limit.
UE transmit timing	The UE should initially set its DPCCH/DPDCH transmit timing to 1024 chips later than the frame timing of the downlink DPCCH/DPDCH. This test verifies the initial timing is within the specified tolerance. Additionally, if the main radio link (the link that provides the reference timing for the UE) is changed during an active set update procedure, the UE has to move its timing to the new reference. However, there are strict rules on how fast the UE can change its timing, with both a minimum and a maximum rate of change, and this test also verifies that the rate of change is within the specification.

12.2.1 Measurements

The UE is expected to make a number of measurements through its receiver both to enable decisions it must make autonomously, and to send back to the network in measurement reports when it has a DCH, to enable the network to make decisions. The types of measurement and the events associated with them are covered in Section 10.2.8 and Table 10.2. The RRM tests cover two aspects of measurement:

- Measurement timing (Figure 12.5): Measurement reports should be sent back to the network within a specified time limit after the event has occurred.
- Measurement accuracy or performance: The UE is required to meet certain accuracy requirements when making measurements, and there are requirements both on the absolute accuracy and on the relative accuracy.

For the various types of measurement, the quantities that can be measured are shown in Table 12.1.

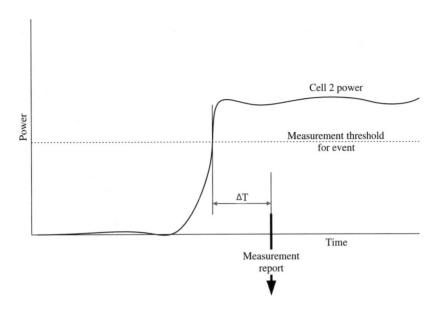

Figure 12.5   Triggering of a measurement report

Table 12.1   Measurements and their purposes

Measurement quantity	Measurement type	Purpose
CPICH RSCP	Intrafrequency Interfrequency	Active set update decisions, hard handover decisions, open-loop power control and pathloss calculations.
CPICH E_c/I_0	Intrafrequency Interfrequency	Cell selection/reselection, active set update decisions and hard handover decisions.
UTRA carrier RSSI	Interfrequency	Interfrequency handover decisions.
GSM carrier RSSI	Inter-RAT	Inter-RAT handover decisions.
Transport channel BLER	Quality	Maintenance of QoS and RB reconfiguration decisions.
UE transmitted power	Internal	Maintenance of QoS.
SFN–CFN observed time difference	All monitored cells	Handover timing for CELL_DCH transitions.
SFN–SFN observed time difference	All monitored cells	There are two types of SFN-SFN measurement:
		• Type 1: based on P-CCPCH timing is used for estimating timing differences between cells for handover in CELL_FACH state.

(*continued overleaf*)

Table 12.1 (*continued*)

Measurement quantity	Measurement type	Purpose
		• Type 2: based on P-CPICH timing is used for estimating UE position for use with LBSs.
UE Rx–Tx time difference	UE internal	There are also two types of Rx–Tx measurement, but the difference between them is quite subtle. Both measure the time difference between receiving the DPCH/F-DPCH on the downlink and transmitting the DPCCH on the uplink, but
		• Type 1: uses only the first path received on one of the DPCHs used for demodulation and is used by the network to compensate for propagation delays due to the distance of the UE.
		• Type 2: uses the first path received on any of the DPCHs in the active set and is used for estimating UE position for LBSs.

12.3 Typical Test Systems

RRM conformance test systems share many parts in common with RF conformance test systems; the main difference being that some of the RRM tests require many more cells than are needed for RF testing. As a result, the RRM systems tend to be built as extensions or add-ons to the RF ones. In the case of the model shown in Figure 12.4, the RRM-specific part is a fourth rack of equipment including two SSs providing the additional cells, added along side the RF test system (compare this with the picture in Figure 8.29).

13

High-Speed Packet Access

13.1 Introduction

High-speed packet access was introduced into the 3GPP specifications in two phases. Release 5 introduced high-speed downlink packet access (HSDPA), with data rates ranging up to 14.4 Mbps (compared to a maximum of 2.048 Mbps in previous releases), and Release 6 introduced it to the uplink in the form of the E-UL, with data rates up to 5.6 Mbps. For the downlink, these data rates are achieved by a combination of higher order modulation and reduced (higher rate) error correction coding, together with an adaptive modulation and coding (AMC) scheme to tailor the selection of both to the available channel conditions. This exploits the fact that the Node B transmitter uses a much smaller dynamic range than the UE, about 20 dB compared to 70 dB. As Node B has to reach the edges of the cell, those UEs closer to Node B can frequently have significant margin over the minimum required SIR to meet the QoS target, and this is effectively a waste of capacity. For the downlink, the standard WCDMA rate adaptation using SF and fast power control is therefore limited by both the need to allocate individual codes to connected UEs, and by the Node B dynamic range. Instead, this is replaced by adaptation through changing the modulation scheme and the channel coding rate. For the uplink, higher order modulation increases the UE complexity considerably, and therefore multiple channel codes are used instead to create parallel physical channels. As well as the higher data rate, radio link latencies in both directions were reduced with the introduction of a 2-ms subframing and a fast retransmission and acknowledgement process, known as hybrid ARQ (HARQ). These are examined in the following sections.

Testing UMTS: Assuring Conformance and Quality of UMTS User Equipment   Dan Fox
© 2008 John Wiley & Sons, Ltd

13.2 Physical Layer

HSDPA is based on the concept of a high-speed common downlink channel shared by all connected UEs in the cell, with scheduling of data for individual UEs. It introduces a number of new physical channels to support this:

- HS-PDSCH, the high-speed downlink-shared channel, which is a shared downlink only channel carrying the user data. This channel uses SF16 only, which for a single code gives a raw symbol rate of 240 kilo symbols per second (ksps). On this channel only, a higher order modulation scheme using 16-point QAM is added to the existing QPSK scheme. This allows four bits per symbol, giving a bit rate of 960 kbps. The rate is further extended by allowing the aggregation of up to 15 channelization codes. This provides a raw bit rate of 14.4 Mbps.
- HS-SCCH, a 60-kbps (SF128) downlink channel which carries control information from the network to the UE, particularly associated with the scheduling of data on the HS-PDSCH.
- HS-DPCCH, a 15-kbps (SF256) uplink channel which carries feedback information relating to the channel quality and layer 2 HARQ acknowledgements.

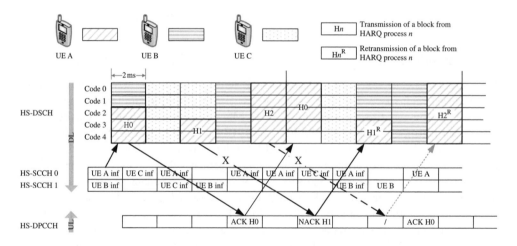

Figure 13.1 Physical channels for HSDPA showing multiple access

Figure 13.1 shows the relationship between the physical channels for HSDPA. UEs can both be time and code multiplexed on the data channel. Within a time slot, each active UE is given a set of codes which Node B will use for transmission. HSDPA only provides data transfer capability on the downlink and on its own cannot support flows that require bidirectional support, such as the SRBs and data transfer using AM-RLC. Therefore, an HS-DSCH connection always has a DCH of some sort associated with it. This can either be provided by an associated DPCH or it can be provided by the uplink E-DCH from Release 6 onwards.

For the E-UL, due to the timing complexities and arbitration overhead of common channels, the use of DCHs is maintained, but a Node B-controlled scheduling system is introduced to minimize the interference from other users. This has the effect of making the combined uplink channels look like one large shared channel. It can be viewed as 'semishared' in that most of the capacity can be directed towards one or a few UEs to allow high speed utilizing the bursty nature of typical packet-switched traffic but still allowing a certain amount of unscheduled use of the channel. The E-UL introduces the following new physical channels:

- E-DPDCH, the enhanced version of the standard uplink DPDCH channel carrying the user data, which adds the use of SF2. The UE can have up to four E-DPDCH on different channel codes. The different combinations are shown in Table 13.1.

Table 13.1 Maximum numbers of physical channels available with HSPA (reproduced by permission of ETSI)

	DPDCH	HS-DPCCH	E-DPDCH	E-DPCCH
Case 1 (HSDPA only)	6	1	—	—
Case 2 (HSDPA, E-DCH and associated R99 DCH)	1	1	2	1
Case 3 (HSDPA and E-DCH)	—	1	4	1

- E-DPCCH, the enhanced control channel associated with the uplink E-DPDCH. Only one E-DPCCH is used and is common to all the E-DPDCH. It carries the enhanced transport format combination indicator (E-TFCI) mapping the TFC used on the E-DPDCH. It also carries the retransmission sequence number (RSN) relating to HARQ retransmissions and a feedback bit known as the Happy Bit, which indicates whether the UE needs more capacity.
- E-HICH, the enhanced HARQ indicator channel, on the downlink at SF128, which carries the HARQ acknowledgements back to the UE.
- E-AGCH, the enhanced absolute grant channel, a common downlink channel at SF256, which carries the absolute grant information to UEs in the cell.
- E-RGCH, the enhanced relative grant channel, a dedicated downlink channel at SF128, which carries grant information relative to previous transmissions from the UE.

For the E-UL, the physical channels are shown in Figure 13.2. The UE can use up to four E-DPDCHs on different channelization codes, as shown in Figure 13.3. Transmissions from multiple UEs are scheduled by Node B using a system of grants, which tell the UE how much

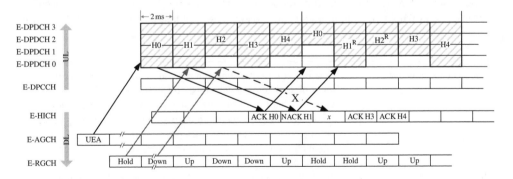

Figure 13.2 Physical channels for the E-UL showing scheduling of capacity (Note, for simplicity only five HARQ processes are shown. For a 2ms TTI there would normally be 8 of these)

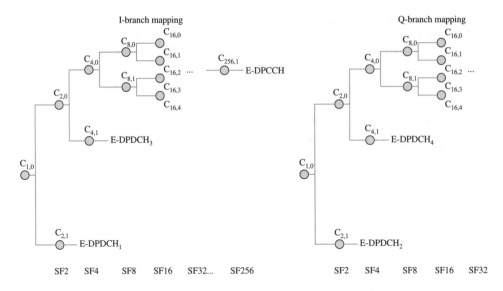

Figure 13.3   Channelization code allocation for the E-UL

power it can use. From this, the UE can estimate the amount of data and select an appropriate transport format and puncturing level. The diagram shows simplistically the number of E-DPDCHs varying as the grant is raised or lowered, but in practice, the granularity is far smaller, with a degree of variation in the amount of redundancy information before the data is spread onto additional E-DPDCHs. The allowable configurations of maximum numbers of channels are shown in Table 13.1.

13.2.1 Hybrid ARQ

The HARQ protocol uses a series of independent processes that transmit blocks of data. Each process transmits a single block of data and then stops until it either receives an acknowledgement, a NACK or if no response is received within a time period. After a configurable number of attempts to retransmit, the process will discard its block and be available for a new block of data. There is also a prioritization mechanism, where data from higher priority streams can pre-empt retransmissions, causing the old data to be discarded early. The following two techniques are used to enhance the likelihood of successful decoding in the receiver:

Incremental redundancy	If the initial transmission cannot be successfully decoded, then it is stored in a buffer, known as the soft buffer as the data is stored before any hard decisions are made as to whether a received bit is a one or zero. In subsequent retransmissions, the puncturing pattern is varied so that different bits from the same data are sent. These are combined with the previous transmissions and a further attempt made to decode the data.
Chase combining	This is very similar to IR, but the same data with the same redundancy information is retransmitted.

The HARQ processing requires very fast decisions, and therefore a part of the MAC functionality has been moved from the RNC to Node B. This is done by the creation of a new entity in the downlink, the MAC-hs, and two new entities in the uplink, the MAC-es and the MAC-e. The MAC-hs and MAC-es are located in Node B and are where the HARQ processes reside. The MAC-e is located in the RNC and performs soft combining of E-UL data from one UE but received by multiple Node Bs. The changes to the network architecture are shown in Figure 13.4.

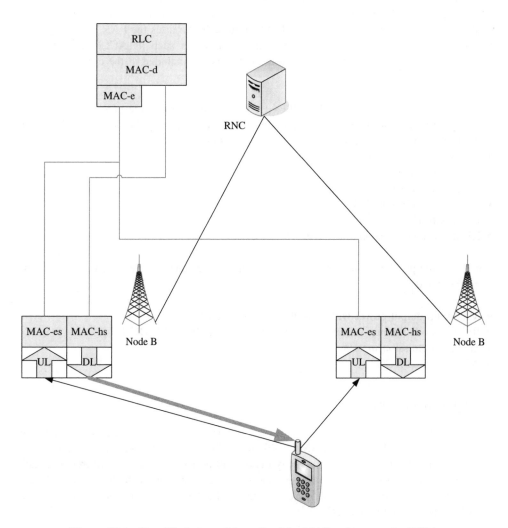

Figure 13.4 Simplified view of the split of the MAC architecture for HSPA

The split of MAC functionality between the RNC and Node B, created an indirect link between the MAC-hs/es and the RLC, so the operation of transport format selection had to be modified. Instead, the MAC-hs and MAC-es entities send a single PDU in a TTI. The size of this PDU varies according to the amount of bandwidth allocated to that UE, and depending on this size, the MAC-hs or MAC-es will fit as many MAC-d PDUs as it can.

13.3 RF Tests for HSPA

13.3.1 Transmitter Tests

For HSDPA, the HS-DPCCH is added to the set of UE-transmitted channels. This is code multiplexed with the DPDCH and DPCCH from the associated DPCH and places some additional physical constraints on the UE transmitter. It increases the PAR, requiring better linearity in the transmitter to meet spectral requirements such as ACLR and spectrum emission mask (SEM). The following additional tests are included in the conformance test specifications:

- Maximum output power with HS-DPCCH. This is split into two tests, one for Release 5 and one for Release 6 and later. The main difference is the way the power conditions for the test are defined with respect to the Release 99 channels.
- Relative code domain power accuracy.
- HS-DPCCH power control.
- Spectrum emission mask with HS-DPCCH.
- ACLR with HS-DPCCH.
- EVM with HS-DPCCH.
- EVM and phase discontinuity with HS-DPCCH.

For E-UL, the problem is magnified as further multicode channels are added. The following tests are added:

- Maximum output power with HS-DPCCH and E-DCH.
- Relative code domain power accuracy with HS-DPCCH and E-DCH.
- Spectrum emission mask with E-DCH.
- ACLR with E-DCH.

13.3.2 Receiver Tests

The receiver characteristic tests are only affected by HSDPA, and a test is added to check the maximum input level with a 16 QAM HS-PDSCH. The receiver performance tests, however, have been significantly extended to verify the overall performance of the different channels. To facilitate these tests, a number of new reference measurement channels have been added. On the downlink, these are termed fixed reference channels (FRCs) and six parameter sets, called H-Set 1 to 6, are defined.

Part III

The Future

14

Future Trends in Testing

14.1 Testing Earlier in the Development Cycle

It is generally recognized that the earlier in the development cycle that testing can begin, the better. Two important trends towards earlier testing emerged during the development of third-generation devices, and these trends are likely to be strengthened in the future.

14.1.1 Virtual Testing

A number of test equipment manufacturers produced virtual or hardware-less test systems aimed at testing the UE protocol stack prior to integration with the hardware. These systems were typically derived from or shared common user interfaces with the full SS, allowing test cases developed in the virtual environment to be easily moved to the full hardware environment. A virtual tester replaces the radio modem and associated channel processing with a simple interface that carries the user data, together with explicit indications of those radio parameters which would normally be implicit in the transmitted signal. For example, the scrambling code is carried implicitly in the physical system through the application of scrambling to the data stream, whereas in the virtual system it would be carried as an explicit parameter 'scrambling_code = $nnnn$'.

While various products exist, the use of virtual testing is not yet widespread. One of the main issues has been that there are no standardized interfaces within the UE. The virtual tester needs to interface at the boundary between layers 1 and 2, and this task is different for each UE development, so that adapting the virtual tester always takes some development effort. As functionality is added to the UE stack, there is also work needed to maintain the UE adaptation. Once the hardware becomes available, the argument for this maintenance effort becomes more marginal as testing can be carried out on the real system; hence the usefulness of a virtual tester becomes less as the development progresses.

Testing UMTS: Assuring Conformance and Quality of UMTS User Equipment   Dan Fox
© 2008 John Wiley & Sons, Ltd

14.1.2 Earlier Use of Conformance Testing

The second trend is in the use of conformance tests. In second-generation systems, the conformance tests came along at the end of the terminal developments and played little role during them. During the development of third-generation terminals, the conformance tests started to emerge quite early in the development process, and as a result basic UE functionality was being tested against conformance tests in parallel with ongoing development. This is highly beneficial, as conformance is far more difficult to achieve if it is left to the end of the development. By testing throughout the development, the accumulation of errors in the UE is avoided, and this leads to a much smoother path through to full conformance. This trend has become more pronounced over recent 3GPP Releases, as the solid foundations on which the conformance tests were built has allowed more rapid development of test cases for new features. Some of the key features for Releases 5 and 6 were available early enough in conformance tests to form a major part of integration and verification of even some of the earliest UE developments. This trend will place a lot of emphasis on the development of conformance tests for future extensions to the standards.

14.2 IMS and Technology Convergence

Current networks consist of two domains with relatively separate network equipment in each domain: one for circuit-switched and one for packet-switched. The data traffic has already largely moved from the circuit-switched domain, so if the voice traffic can also be moved into the PS domain, then the operation, maintenance and provisioning of the network will become easier. The core technology needed for this is voice over IP (VoIP), which together with a number of associated services, such as presence indication, is embraced by the IP multimedia subsystem, or IMS. This refers to a collection of functional entities and protocols aimed at providing the same wide variety of multimedia over fixed and cellular networks that can currently be experienced on the Internet. While the networks today generally have access points to the Internet via the GGSN and can set up transparent pipes to such multimedia services, the goal or IMS is to integrate this into the network in such a way that these services can be provided at a level of quality and security comparable to current operator-provided cellular services.

IMS was first introduced into the 3GPP specifications in 2003 (Release 5). Similar concepts were already under development for the fixed network; in Europe, this work was driven by the ETSI TIPHON project (Telecommunications and Internet Protocol Harmonization Over Networks). In 2003, TIPHON was merged with another ETSI committee (SPAN) to form a committee, titled 'Telecommunications and Internet converged Services and Protocols for Advanced Networks', or TISPAN for short, with a much broader scope of making IMS access technology independent. Both 3GPP (various committees) and TISPAN are now active in progressing the IMS suite of specifications, with 3GPP focussing mainly on the cellular network aspects.

From the user's perspective, IMS will provide seamless communication between both people and applications regardless of whether they are mobile, PC-based or fixed terminals. From the operator's perspective, IMS allows a migration from a very centralized network based on large, expensive switches to a more distributed network using lower cost components. The architecture for IMS in the mobile network is shown in Figure 14.1. Whilst

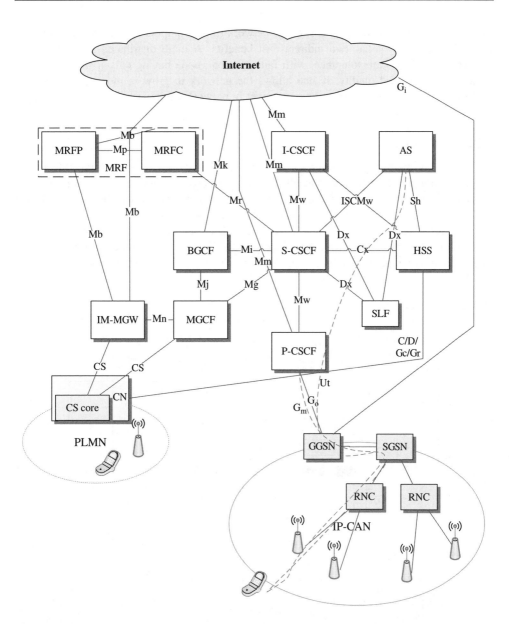

Figure 14.1 Network architecture for IMS. *Key*: AS, application server; BGCF, breakout gateway control function; CN, core network; CS, circuit switched; CSCF, call session control function; HSS, home subscriber server; IBCF, interconnection border control function; I-CSCF, interrogating CSCF; IM, IP multimedia; IM-MGW, IM media gateway; IMS, IP multimedia core network subsystem; IP-CAN, IP-connectivity access network; MGCF, media gateway control function; MGF, media gateway function; P-CSCF, proxy CSCF; PLMN, Public Land Mobile Network; S-CSCF, serving CSCF; SLF, subscription locator function

this in itself does not reduce the cost of the network (more components are needed to provide the same capacity), it has two indirect cost benefits. A more distributed network has an element of built-in fault tolerance, with failure in one node having only a small effect on network capacity and quality. It also allows the network to grow in more gradual stages, reducing the speculative element of investment in network equipment.

14.2.1 Testing IMS

IMS raises an interesting challenge for testing methodology in two ways. First, IMS has its heritage in the world of IP, and typically testing of IP products and services has been much less formal than for the telecoms arena. The industry has to resolve whether testing of IMS will follow a more typical IP testing route or the more formal conformance approach. In fact, the indications are that it will adopt a 'best of both worlds' approach, with a combination of conformance test suites from the telecoms side and extensive more empirically based interoperability testing with joint testing sessions ('Test-fests' and 'Plug-fests') more typical of IP testing.

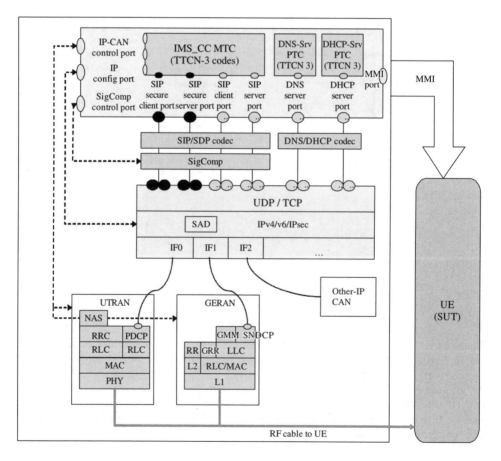

Figure 14.2 Architecture of the 3GPP IMS test suite for SIP-based call control (reproduced by permission of ETSI)

The second question that arises is how IMS development testing will be done. In the past, a clear differentiation has appeared between testing of applications and testing of core signalling protocols. Applications tend to be tested much more in isolation. Typically, the task is to verify the functionality of the application and often the bulk of this testing can even be done without the UE. Particularly for IP-based applications, if the application is structured independently of the transport medium, it can be tested using PCs and Ethernet connections. In theory, core signalling protocols could also be tested the same way, but they tend not to be as this would miss a lot of potential problems through service interactions and environment interactions. Today, IMS or similar IP applications included on current generation UEs tend to be mainly tested as applications, that is, in isolation using simple radio links and often with substantial testing taking place without the target platform. However, as IMS applications transition to become the core signalling protocols of the UE, the industry needs to decide whether the test methodology should also transition from pure application testing towards more complete protocol testing.

Based on previous work from the ETSI TISPAN committee, IMS conformance tests are already becoming available from 3GPP. The architecture of these tests (Figure 14.2) sits somewhere between the two methodologies. It uses controllable emulations for the underlying protocols of the IP-CAN but mainly tests the IMS SIP protocol in isolation. As more IMS protocols are added, this solution could evolve in either direction.

14.3 Evolving Testing Technologies

14.3.1 TTCN 3

TTCN 3 was standardized by the ETSI methodologies for testing and specification (MTS) committee, who were responsible for the maintenance of the TTCN 2 specifications. Work on TTCN 3 started in 1997, and the new standard, published in October 2000, was a fairly extensive overhaul of the version 2 specifications and language, even to the point that while the initials 'TTCN' were kept, the meaning was changed from 'Tree and Tabular Combined Notation' to 'Test and Testing Control Notation'. TTCN 3 builds on some of the strengths of TTCN 2 but adds significant new functionality, a fair amount of which comes from many years of experience of practical usage both in conformance testing and in development testing. One of the principal changes introduced by TTCN 3 is that the core language is much more programmatic in nature and has now look and feel of a regular programming language. The TTCN 2 tables were difficult to manipulate and edit and required specialized editing tools, and this is largely addressed by the new text-based programming language approach. In addition, there are a host of new features and the strengthening of many existing features:

- In addition to PIXITs and PICS, that allow test suite configuration, dynamic test configurations have been introduced. These provide greater flexibility in mapping the test suite to a specific test platform implementation.
- TTCN 2 only allowed for asynchronous communication with the unit under test, TTCN 3 introduces support for synchronous communication.
- Support has been added for testing distributed systems.

- Some attempt was made in TTCN 2 to standardize the interfaces to the test system, but this was largely not rigorous enough and not widely adopted, TTCN 3 adds a formally defined test system interface.
- A broader set of data types, including improved text string matching using regular expressions, have been added.
- Particularly driven by 3GPP, abstract syntax notation one (ASN.1) has become important for the specification of signalling, with close ties to codec implementation. TTCN 2 offered interworking with ASN.1, but for TTCN 3, a much greater harmonization has been achieved.
- With the convergence with IP technologies, the language has been extended to integrate other language types, such as XML.
- Support for parallel test components has been greatly enhanced. These allow much better structuring of tests of very large systems and are particularly well suited to IP technologies where there are many functional entities and associated protocols distributed around the network.
- TTCN 3 also offers much better support for modular test suite design, again an increasingly important factor as the complexity of test suites increases.

Over the next few years, the industry faces a challenge as testing moves from TTCN 2 to TTCN 3. Already this transition has started. In 3GPP, new tests based essentially on the existing standards have to still be provided in TTCN 2 as the task of porting the legacy test suites would be very large. New Release 7 features will still be tested using TTCN 2, and this will probably be the same for future WCDMA extensions for the foreseeable future. Where new standards are being introduced and the legacy benefit of the existing WCDMA test suites is lower, the tests are generally being provided in TTCN 3. This is in fact the case for IMS (as mentioned in the previous section) and for the planned 3GPP LTE.

14.3.2 Graphical Tools

Standardized testing tools such as TTCN 2 and TTCN 3 are very much focussed on providing a formal definition of the test and the expected behaviour that allows the unambiguous implementation of tests. This allows the definition of tests which can run in a similar enough way on different test platforms such that the results of testing are valid independent of the test platform used. However, this platform independence in practice needs more than just the language and usually requires significant cooperation within the industry to achieve. For broadly available conformance test suites, the effort to do this is worthwhile and has indeed been practically realized for UMTS. However, for the development, integration and verification testing carried out within a UE developer, the goal of platform independence is not practical, and the overhead in terms of the lack of ease of use that the formal languages have is not always justifiable. TTCN test suites need careful structuring to make them maintainable, and it takes significant effort to develop high-quality tests. As the complexity of modern systems increases, the amount of testing required, and therefore the number of test cases, also increases. There is a growing need, particularly within UE developers and network operators, to supplement the conformance tests with large numbers of additional, broader tests. As a result, there is still a demand for easier to use test systems, and a number of more graphical approaches have emerged over the last few years.

14.3.2.1 Test Development Automation

One of the key benefits these tools offer is the automation of the SS configuration. Formal tools place the configuration of the SS in the hands of the test author. In fact, formalizing this configuration is a large part of the way in which different test platforms are made to behave in the same way. However, creating the SS configuration and matching it to the commands being sent to the UE constitutes a large part of the work of writing test cases. Modern test tools allow the test developer to work in terms of higher level signalling procedures and can extrapolate the SS configuration from the parameters that the test author applies at the procedure level. Such a tool can provide the flexibility of a TTCN-based system (e.g. mismatched configurations to simulate failure scenarios) whilst greatly accelerating the development of test cases. Figure 14.3 shows a new generation graphical test development tool which offers both automation of SS configuration and the ability for an advanced user to override the automation to create specialized scenarios.

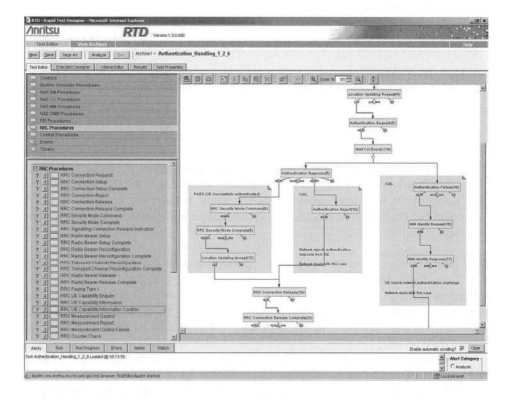

Figure 14.3   Screen shot from a graphical test development tool (Anritsu Rapid Test Designer, reproduced by permission of Anritsu)

14.4 Future Cellular Network Technologies

Whist the development of IMS offers a revolution in the CN, the access network is also evolving in two directions. The current architecture is moving towards the use of fewer types of node, distributing the processing more widely in the network. From a UMTS perspective this is covered under the title of system architecture evolution (SAE) which introduces a number of concepts:

- An evolved packet core (EPC) which replaces the current SGSN/GGSN structure.
- The removal of the RNC and transfer of the RNC's functionality towards the edge of the network to create an evolved Node B (eNB).
- The use of an all IP backbone with the eNBs all interconnected in an Intranet style network.

Rather independent of this in technical terms, but also part of the future development of UMTS is the introduction of a new air interface technology known as 3GPP LTE. This is based on the use of orthogonal frequency division multiple access (OFDMA) for the downlink, and a closely related technology: single carrier frequency duplex multiple access (SC-FDMA) for the uplink. The goal is to introduce a capacity of up to 300 Mbps in the downlink and 75 Mbps in the uplink.

In parallel with this, the WCDMA-based air interface continues to evolve. Release 7 has introduced a number of new features to further extend the performance of HSDPA and E-UL:

- Multiple input–multiple output (MIMO) antenna technology, where the concept of transmit diversity is extended such that if the channel quality is good enough, different data can be transmitted on each antenna, coded in such a way that the UE can recover both streams by combining the images on each receive antenna. This potentially doubles the data rate, providing up to 28.8 Mbps on the downlink. This technology is also central to LTE.
- Even higher orders of modulation, moving the downlink up to 64 QAM and the uplink to 16 QAM, providing respectively 50 and 100 % more performance (21.6 and 11.5 Mbps).
- Continuous packet connectivity (CPC), a collection of features aimed mainly at enhancing the ability of the mobile to remain connected whilst not actively transmitting data by providing ways to reduce the power consumption when inactive and speed up the signalling to return to activity.

All of these new developments will have a significant impact on testing. The introduction of LTE requires a completely different baseband technology in the SS, together with a significantly different set of test challenges both in the RF domain and for the signalling protocols. The higher speeds and other RF changes coming from the evolution of HSPA technology will require enhancements to existing test equipment together with many new RF tests. Both developments are also placing much more emphasis on data throughput. In the past, this has been of secondary importance to the operation of the underlying protocols, but high 'headline' data rates are acting as a stimulus for applications that need mobile broadband performance, such as video streaming and mobile TV-on-demand. The ability of SSs to support throughput and Quality of Service testing is likely to be one of the biggest areas of growth over the coming years.

Appendix

Tree and Tabular Combined Notation

A.1 Introduction to TTCN

In 1984, the ISO started work on a standardized approach to testing based on their seven-layer protocol model. This work was done in partnership with the ITU and resulted in an identical specification being standardized by both organizations; for the ISO this became ISO/IEC 9646, and for ITU it became the X.290 series. The specification came in five parts, looking at various aspects of testing, and included a formal language for describing tests, and these are described in Table A.1. TTCN itself is described in part 3 of the series. This chapter provides a brief introduction to the TTCN language. However, the intention is to familiarize the reader with the basic concepts and the key elements of the language. This is not an exhaustive description but should provide enough background to allow the user to make sense of the specification.

A.1.1 Basic Concepts

The ISO/IEC 9646 approach, driven by the structuring of protocols into the layered model, identifies that protocol implementations have an upper and a lower bound. Therefore in theory, a protocol implementation can be tested by applying the appropriate stimuli at each boundary and checking the responses and actions. If the protocol layers are properly formed then this should also apply if the protocols are stacked and the upper bound is the top of the upper layer and the lower bound the bottom of the lower layer (Figure A.1).

This model can be extended to fit any number of protocol layers in the implementation under test (IUT), even up to a complete protocol stack implementation. The model needs a

Testing UMTS: Assuring Conformance and Quality of UMTS User Equipment   Dan Fox
© 2008 John Wiley & Sons, Ltd

Table A.1 The parts of the TTCN specification

Specification	Description	ITU-T reference
ISO/IEC 9646-1	Information technology – Open Systems Interconnection – Conformance testing methodology and framework – Part 1: General concepts.	X.290
ISO/IEC 9646-2	Information technology – Open Systems Interconnection – Conformance testing methodology and framework – Part 2: ATS specification.	X.291
ISO/IEC 9646-3	Information technology – Open Systems Interconnection – Conformance testing methodology and framework – Part 3: The Tree and Tabular Combined Notation (TTCN)	X.292
ISO/IEC 9646-4	1994, Information technology – Open Systems Interconnection – Conformance testing methodology and framework – Part 4: Test realization.	X.293
ISO/IEC 9646-5	Information technology – Open Systems Interconnection – Conformance testing methodology and framework – Part 5: Requirements on test laboratories and clients for the conformance assessment process.	X.294

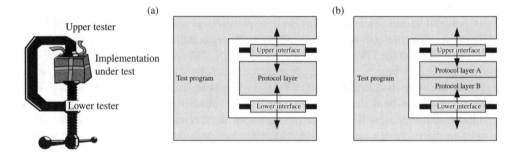

Figure A.1 The concept of upper and lower testers in TTCN. (a) Testing a layer at its upper and lower boundaries. (b) Extending the model for multilayer protocols

little extension to cope with the realities of getting signals into a real implementation and is usually shown as in Figure A.2. The reasons for these extensions are as follows:

- When the IUT becomes many-layered, by definition, the interface at the top will be high level. A few commands will equate to a large amount of work down at the lower interface. Therefore it makes sense to separate the two parts, at least logically if not in physical implementation, so that the operations at the top and bottom of the stack can each be treated in their own context.
- Usually complete stacks end up with some physical realization of the lowest layer, such as a radio or an electrical transceiver. This is not actually part of the test program, but still the test program needs to interface to the IUT through it. Therefore, it is shown as a logically separate entity: the service provider.

- Finally, there needs to be some coordination between the upper and the lower testers. When an action is triggered by the upper tester, it will result in some consequential activity with the lower tester, and these need to be synchronized and managed together.

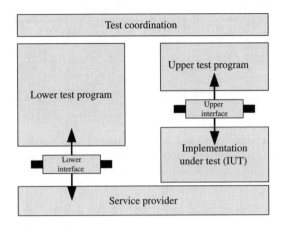

Figure A.2   Generic structure of a TTCN tester

We can extend this model further. Many complex systems actually have several interacting protocols running, that is, parallel layered protocols in the IUT. The whole concept can be extended to have multiple test programs, one for each protocol, running in parallel and communicating with their respective protocol entities in the IUT. This is indeed the case for UMTS and other mobile communications standards, but although TTCN does have capability to deal with parallel test programs, in practice parallel testing is difficult, and its use was avoided by 3GPP.

A.1.2 Test Methodology

In theory, the simplest approach is to break the testing task into two: the first is to test each protocol layer in isolation, and to not allow the stacking of layers, and the second is to test the full protocol stack as a single entity to make sure the layers work together correctly. These are known respectively as horizontal and vertical testing. However, there is a problem to overcome. In practice, it is very difficult to get inside an IUT, which is usually software within a physical device, and separate out the layers cleanly. In some cases, test loops and test interfaces can be added, but there is a limit to how much overhead an implementation can realistically be expected to tolerate just for the purposes of conformance testing, and hence most interlayer interfaces are closed to the test program. The horizontal/vertical approach is still broadly followed, but the separation of the layers is a compromise towards practicality.

This is shown in Figure A.3, which relates this approach to UMTS. The full stack is used to get the device into a state where the target layer under can be tested thoroughly. There is no universal formula for this, but often test loops can be used to verify behaviour within the layer. Where there are layers above and below, the tests are designed to use the minimum functionality within them, but this does represent the main weakness of this approach, in

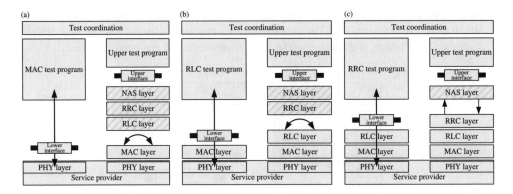

Figure A.3 The TTCN tester as applied to UMTS layer 2 and 3 testing. (a) Practical approach to testing by layers MAC is tested using test loops. (b) RLC is tested using the MAC. (c) Testing progresses up the stack layer by layer. Where loops cannot be used minimum functionality in the layers above is assumed

that it does rely on the layers apart from the one being tested all functioning correctly. For conformance testing, this is an acceptable assumption. If the other layers do not work properly, then the test will fail, but this is a fair result as the other layers are clearly not conformant.

Basically, the architecture of the 3GPP protocol test suites closely follows this model, as can be seen from Figure A.4. The main difference is that the upper test interface is at a very high level and quite straightforward to drive. The commands for the upper tester

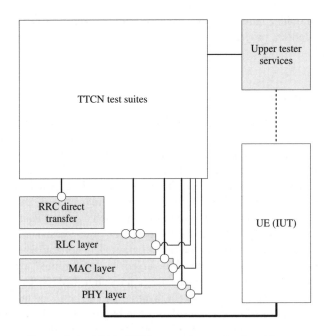

Figure A.4 3GPPs TTCN test suite top level architecture

are therefore embedded in the TTCN test suites along with the lower test program. This allows for easy coordination between the two. The upper tester uses external services to communicate with the UE, and these (along with the other shaded components) are provided by the test equipment manufacturer.

A.1.3 Dynamic Behaviour

To understand the structure of TTCN, we first need to look at how TTCN works. In its simplest form, testing involves sending signals or stimuli into the IUT and looking at the signals that come back from it. Following the principle of noninvasive testing, these stimuli are usually signalling messages, PDUs (for the lower layers) or some kind of physical action on the unit itself (e.g. pressing a button or switching the power on). Applying a stimulus and looking at the result will only test the simplest of interactions, so a more detailed test will need to apply a sequence of stimuli. The sequence can also depend on some of the results received on the way. This is shown in Figure A.5.

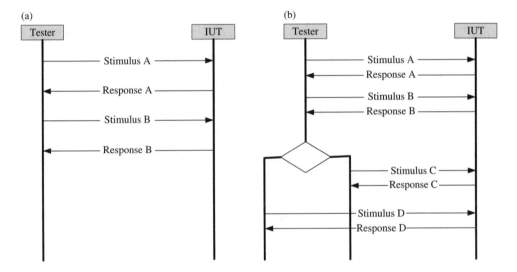

Figure A.5   Applying a stimulus to the IUT and receiving responses. (a) Basic testing: applying a sequence of stimuli and examining the responses. (b) Conditional testing: depending on the responses in A and/or B, either stimulus C or D are sent

   Test cases can have branches that themselves branch and this can run to many levels. It is easy to visualize this as a tree, where each node of the tree is a decision point that spawns two or more branches (Figure A.6). This is the 'tree' part of TTCN. By drawing the tree on its side, it is easy to see the relationship in the way the tree is represented in TTCN. This is also known as the dynamic behaviour. Statements written at the same level of indentation are the alternative branches of behaviour resulting from decisions made at the previous level. Whilst this looks very logical, it is one of the main challenges for newcomers to TTCN. In most other computer languages, indentations are cosmetic only, but in TTCN they are

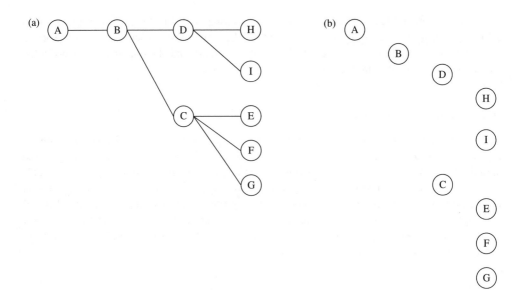

Figure A.6 Simplified view of the tree structure for TTCN dynamic behaviour. (a) Tree structure to describe test flow in TTCN. (b) Lay out of test case statements using indentation to represent the tree

operational and affect the flow of the program. Furthermore, each step in the program takes you one further indent along (you cannot take both paths of a choice), so a long program can have a high degree of indentation. This makes it difficult to display, read and follow the structure of the tree. TTCN does allow hierarchy within programs, through a structure similar to a subroutine (called a test step), and an experienced programmer will use test steps to limit the level of indentation and keep the trees readable.

A.1.4 Constraints

So far we have talked about sending stimuli and receiving responses, but for protocol testing most commonly these are signalling messages, often containing large amounts of information. A message has a structure, which is described in the relevant signalling specification, and is generic for all messages of the same type. Then a message also carries information which is specific to the purpose behind the sending of the message in this particular instance. The message structure is a global property of that message type and in TTCN is handled by a 'declaration'. In the message instance, concrete values have to be assigned to the fields within the message, and this is done with a construction called a 'constraint'. These are shown in Figure A.7 using the NAS Call Setup message as an example.

Another, often confusing, part of TTCN is that constraints are defined outside of the dynamic behaviour. This is very different from C (or other mainstream languages), where values tend to be applied to variables close to the point where they are used. This separation of message value setting does have a purpose though. Messages are sent to perform specific tasks. In the example given, the task is to initiate the setting up of a particular type of call. While there are many information fields, there are practical limitations on the number of

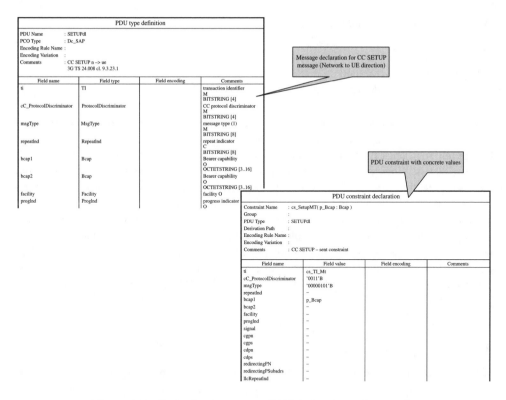

Figure A.7   Example of a message PDU declaration and constraint

different tasks this message will be used for. By thinking of messages in terms of the tasks they will be used for, the benefits of separating this from the dynamic behaviour through the use of the constraint mechanism can be seen more clearly. A constraint effectively 'constrains' a message to a specific task. Constraints are themselves parameterized in that some of the values assigned to message fields can be left blank when the constraint is created, and then filled in during the dynamic behaviour. This means that how messages are separated into various tasks is to a large degree up to the test designer, and getting this right is part of the art of good test suite design. If the constraints have too many parameters their meaning is less clear, the readability and maintainability of the test suite is reduced and there is more scope for error. However, if there are too few parameters, then the library of constraints becomes very large and again readability and maintainability are lost. In the ETSI TTCN style guide (ETR 141), which has a lot of very useful guidance on structuring large TTCN test suites, a maximum of five parameters is recommended.

One approach is to categorize messages according to their purposes, but this categorization can become quite sophisticated as TTCN allows for a hierarchy. For example, we can first break down our call set-up into:

- Call set-up for speech
- Call set-up for circuit switched data (modem).

We can then further decompose the modem part into:

- Analogue modem
- Digital 64 kbps ISDN modem
- Digital 56 kbps modem for restricted networks (RDI).

This hierarchy is supported through the use of constructs known as base and modified constraint. The base constraint will contain the general parameters that differentiate between the speech and data calls. The modified constraints have a parent from which they inherit all their values. However, values specified in the modified constraint override those from the parent. In this example, we would define three modified constraints for the different modem types, each having the data call base constraint as its parent. Modified constraints need to be treated with some caution. If the structure is well thought through then they are very useful, but it can easily become difficult to follow which values are actually being used when a message is sent. To add further complication, the hierarchy can be arbitrarily deep as a modified constraint can have another modified constraint as its parent, but deep hierarchies are notoriously difficult to read and maintain, and the ETSI style guide recommends not going beyond two levels.

A.1.5 Receive Constraints

We have looked at setting specific values into a message through using constraints. This obviously applies when sending messages, but receiving needs a different approach. The concept of a message having a purpose is still applicable, but now we are not setting values. Rather, we are expecting to receive values. Using a constraint in a receive command basically says: 'when I receive something I expect it to contain the following values'. When sending a message, concrete values for all the message fields are needed as it is not possible to send indeterminate fields. This includes the case of optional fields, where if the field is to be omitted then this must be specified, and TTCN has a way of indicating that a field is to be omitted. When receiving, however, there are a number of other possibilities that need to be considered:

- A field should contain one of the values in a list, or
- It should not contain one of the values in the list
- The field should not be present in the message (omitted)
- The field should be present, but we do not care what value it has, or
- We do not care if the field is present or not, and if it is present, we do not care what value it has
- The value in the field should lie within a specific range
- The use of wildcards to match the content of a field, in particular the standard regular expression matching wildcards of '*' and '?', where '*' will match any number of succeeding elements with any value and '?' will only match the next element with any value
- The field can contain specified elements or values, but in any sequence
- A field should be of a specified length.

TTCN has syntax that allows the creation of receive constraints that cater for any of these possibilities.

At this point, it is worth looking at the various data types that TTCN supports. These can be categorized as basic types and structured types, and are shown in Table A.2.

Table A.2   TTCN simple data types

Type	Description
INTEGER	This is simply defined as a positive or negative whole number, including zero. TTCN does not place bounds on integers, but of course, real implementations of TTCN tools have to in order to map onto the finite constraints of processors and memory. Generally, TTCN tools map this to a 32-bit limit, including the sign.
BOOLEAN	Can have a value of TRUE or FALSE.
BITSTRING	This is binary data in a sequence of '1's and '0's followed by the characters 'B. For example 0101'B represents the value five.
HEXSTRING	This is hexadecimal data represented by a sequence of characters from '0' to 'F' followed by the characters 'H. For example 03F'H represents the value 63.
OCTETSTRING	This is a variant of the HEXSTRING where the data sequence must be pairs of hex digits (representing eight bits per pair). It is differentiated using the characters 'O after the pairs of digits. For example, the value 63 would be represented as either 3F'O or 003F'O.
R_TYPE	This means result type and is used for handling of test results. It can have the value pass, fail, inconc (for inconclusive) or none.
Character string types	TTCN has a family of character strings, all consisting of one or more characters enclosed within double quotation marks. For example, 'this is a string'. The difference between each family member is in the permissible character set that can be used. The most common type is the IA5 character set (International Alphabet No. 5) defined by ITU-T T.50, which is the almost identical to the standard American 8-bit ASCII character set. The full set of supported string types is – NumericString – PrintableString – TeletexString – T61String – VideotexString – VisibleString – ISO646String – IA5String – GraphicString – GeneralString – BMPString – UniversalString The formal definition of these types is borrowed from ISO/IEC 8824-1, which is the ASN.1 specification.

These basic types are known as 'built-in types', and they can be used both to generate user-defined variations called simple types and they can be combined together in various ways to make more complicated types, known as structured types.

The main use of user-defined simple types is to constrain the range of values allowed in a field. For example, if we want to represent the UMTS radio bearer identity, which can be a whole number between 0 and 31, we can define this within variables and message fields as INTEGER. For example:

> *myRadioBearerId*   INTEGER

However, in a constraint or assignment, we can set this to any whole number. If we assign this to 32 or more, then clearly we have done something wrong. TTCN allows us to constrain the definition:

> *myRadioBearerId*   INTEGER(0..31)

which now has the advantage that assigning 32 will generate an error and we can easily identify the problem. We are likely to use radio bearer identities all over the place, though, and constraining the definition each time is prone to error. Instead, we can define a simple type and then use that

> Type definition: radioBearerId_type   INTEGER(0..31)

> Variable declaration: *myRadioBearerId* radioBearerId_type

A range of constructs are available to specify simple types, including lists of values and strings with fixed or constrained (minimum and maximum) lengths.

Structure types are built by grouping together simple and built-in types. The simplest way of doing this is to have a sequence of types. For example:

Structured type definition
Type name: MyStructuredType

Element name	Type definition	Comments
Element1	INTEGER	
Element2	radioBearerId_type	
Element3	IA5String[10]	

This creates a three-element structure, where the first element is an unbounded integer, the second an integer from 0 to 31 and the third a character string of fixed length 10 characters. Structured types can be nested, in that an element of a structured type can itself be a structured type. Structured type declarations can also be done using ASN.1 syntax (ISO/IEC 8824-1), which supports a rich set of structures, but some caution should be taken as not all ASN.1 types are supported in the commercial TTCN tools.

In addition to constraints and specific data types, TTCN supports variables and constants, with scoping, and the normal range of arithmetical, relational and Boolean operators. This provides a great deal of flexibility and power. Fields from messages can be copied into or assigned values from variables. They can be operated on mathematically or used in computational algorithms.

A.1.6 Mapping to a Real Test System

As can be seen from Figure A.4, which shows the architecture of the test suites for UMTS, a complex system will usually have many different interfaces over which messages can be sent. TTCN provides a mechanism for identifying each interface, and then within an interface it is possible to define which types of messages can be sent. This provides a level of consistency checking in that the tool is given some rules to identify whether the wrong message is sent on an interface. The interface mechanisms are available:

- Point of control and observation (PCO): This is used to define the interfaces with the underlying test system itself. As such, PCOs cannot be arbitrarily declared. They have to map to something within the underlying service provider layer provided by the test system. PCOs can be either upper tester or lower tester.
- Communication point (CP): This is used for communication between parallel test components, both with each other and with the master test component, in test suites that employ parallel TTCN. As the UMTS conformance tests do not use parallel TTCN, these are mentioned only for completeness.

PCOs can carry both control information, such as configuration commands for the various entities within the test system, and data for transmission to the IUT. The communication takes place using a container called an Abstract Service Primitive (ASP). This is best illustrated with an example from the UMTS test suites. The test cases need to configure the physical layer of the test system according to the needs of the test. The physical layer has a single PCO defined as follows:

PCO declarations			
PCO name	Type	Role	Comments
CPHY	CSAP	LT	Control physical layer

Associated with the CPHY PCO are a number of ASPs which carry the configuration commands for the physical layer, for example:

- CPHY_Cell_Config_REQ, which configures some of the basic cell parameters, such as its frequency, scrambling code and timing
- CHPY_RL_Setup_REQ, which configures a physical channel on the radio link, such as a DPCH or CPICH
- CPHY_TrCH_Config_REQ, which sets the transport channel parameters (TTI, TFS and so on) associated with a physical channel.

There are many more of these (refer to TS 34.123-3 for a more complete list). The ASPs in the example are sent from the TTCN to the physical layer, but there are also examples that travel in the opposite direction, such as CPHY_Out_of_Sync_IND, which tells the test case when the physical layer synchronization is lost. Control ASPs, such as the ones in the example, usually have well-defined structures that do not change each time the ASP is sent or received. The data ASPs, in contrast, need more flexibility. There are a large number of air interface messages, and defining an ASP for each one would be wasteful. TTCN provides a construct called a PDU for this purpose. A PDU definition can be created for each message or data structure exchanged with the IUT. A data ASP will typically contain a number of elements containing routing or similar information used or generated by the tester's lower layers in sending or receiving the data, such as a channel identifier or transmission priority, followed by an element of type PDU. This element can then be set to contain any of the declared PDUs. In the example in the G_LLC_UNITDATA_REQ table below, the 'msg' element of the ASP can be used to send any PDU declared in the test suite; in this case, this is the set of NAS air interface messages.

ASP name	G_LLC_UNITDATA_REQ	
PCO type	G_DSAP	
Comments	The ASP is used to send L3 PDU to the UE/MS in LLC unconfirmed transmission.	

Parameter name	Parameter type	Comments
LLMEId	LLMEId	
Tlli	TLLI	
sAPI	SAPI	
protectMode	BITSTRING[1]	0 – unprotected
		1 – protected
cipherMode	BITSTRING[1]	0 – sent without encryption
		1 – sent with encryption
Msg	PDU	L3 PDU

A.1.7 TTCN Statements

Like any conventional computer language, the program itself consists of a sequence of statements. In TTCN, there are three types of statements:

- Events: These either generate stimuli (send events) or handle responses or timeouts (receive events).
- Constructs: These manage program flow.
- Pseudoevents: These are actions that do not directly affect the IUT or the result of the test and include assignments to variables and starting of timers.

The two key events that interact with the IUT are send, which is denoted by an exclamation mark '!', and receive, which is denoted by a question mark '?'. The formal definition is

"*PCO Identifier*" "!" or "?" "*ASP Identifier*" optionally followed by " ("*Assignments*")"

For example:

Test case dynamic behaviour
Test case name : Example_1

No.	Label	Behaviour description	Constraint ref	Verdict	Comments
1		Dc ! RRC_DataReq	Setup_msg		
2		Dc ? RRC_DataInd	Alerting_msg		

This sends an ASP with a SETUP message to the Dc PCO and waits to receive an ASP containing an ALERTING message from the same PCO. For a send event, the optional assignments can set values in the ASP, but the send event can also be associated with a constraint using the Constraint Ref column, which is a much better way to set the concrete values that the ASP will carry. For a receive event, the assignments are often useful for storing parts of the incoming PDU in variables for use later in the test flow.

TTCN also has constructs that make it easy to place timers around events, as frequently the specifications require responses to be made within certain time limits. Timers are started with the START command and can also be cancelled or the current value read using CANCEL or READTIMER, respectively. Timer values (the length of time they run) can be specified when the timer is started, but TTCN provides a better way of doing this when the timer is specified. This matches closer to the way timers are used in protocol specifications.

Test case dynamic behaviour
Test case name : Example_2

No.	Label	Behaviour description	Constraint reference	Verdict	Comments
1		START t303			
2		Dc ! RRC_DataReq	Setup_msg	P	
3		Dc ? RRC_DataInd	Alerting_msg	F	
4		*More statements...*			
5		? TIMEOUT t303			

In this example (simplified from the real protocol), the timer t303 is started, the Setup message sent and then two alternatives are provided. If the Alerting message is received first then the statements at line 4 are executed, but if the timer t303 expires first then they are not executed. This example also introduces the Verdict column, where pass, fail or inconclusive verdicts can be assigned if the alternative is taken. These can also be abbreviated to the letters P, F and I, respectively. In this case, if the Alerting is received first the verdict is set to pass, and if the timeout occurs then it is set to fail.

A.1.8 TTCN Execution

We have now covered enough of the basics to look at how TTCN executes. Program execution proceeds from top to bottom and to the next level of indentation, one line at a time, unless otherwise directed by one of the flow control statements. Conceptually, any statements on a single line are executed at the same time. If the statement is an assignment, program flow change, pseudoevent or send event, it is immediately performed. If it is a receive event, then the program will wait until the event occurs. If there are alternative events specified (at the same level of indentation), then the program will wait until one of the specified events occurs, and then execution proceeds from the line immediately following the matched event, provided it has a higher indentation level. Execution never proceeds to a line with a lesser indent unless directed by a flow control statement. In the example below, supposing that of the three alternatives B, E and H following step A, alternative E is selected, then the flow of execution will run A, E, F, G and then stop at that point. Although there are lines below, to get to them would require going back to a lesser indentation level:

Test case dynamic behaviour
Test case name : Example_3

No.	Label	Behaviour description	Constraint ref	Verdict	Comments
1		!A			
2		?B			
3		!C			
4		?D			
5		?E			
6		!F			
7		?G			
8		?H			
9		!I			
10		?J			
11		K			

The handling of receive events deserves a little more explanation. It can be seen that there is a possibility for uncertainty and ambiguity around exactly when and in what order events arrive. For instance, in the previous example, what happens if during the execution of statement A, events B and E arrive very close together? The behaviour in these types of situations is well defined by the TTCN specification. For each PCO, the TTCN framework maintains a queue. Events that occur on the PCO are placed in the queue in the order they are received. A separate queue is kept for timeouts. When a set of alternatives are to be evaluated, a snapshot of the status of all the PCO queues is taken. That is, once the evaluation of the alternatives is started, any events that arrive are not considered for that evaluation. If no events received at the time of the snapshot match any of the alternatives or if no events have arrived yet, another snapshot is taken and another evaluation performed. This process continues until an event matches an alternative. This process still leaves a small possibility that in between evaluation loops two messages could arrive at different PCOs. In this case, the evaluations are performed in top to bottom order, and the first alternative to match is taken. One further important rule is that events remain in the PCO queue until they are matched. Considering the previous example, if at the point where we are waiting for events B, E or H, assuming they are all on the same PCO, suppose that event D is received first. This will go to the head of the PCO queue but will not match any of the alternatives. Now the program flow cannot move on, and in fact, even if an event B, E or H is subsequently received, it will remain stuck in the queue behind the event D. Consequently, it can be quite easy to write test programs that get stuck in endless loops without completing or returning a verdict. Apart for the timer mechanism, which can be used to put guard timers around receive events, TTCN also provides two other mechanisms that help handle unexpected behaviours: the OTHERWISE statement and the concept of default behaviours.

The OTHERWISE statement is used essentially to terminate a list of alternatives. It will successfully match any event not specified in the alternatives. For instance, if the previous example is extended (Example_4) to add an OTHERWISE as a fourth alternative, in this case if an event D is received, processing will continue from line 13.

Test case dynamic behaviour
Test case name: Example_4

No.	Label	Behaviour description	Constraint reference	Verdict	Comments
1		!A			
2		?B			
3		!C			
4		?D			
5		?E			
6		!F			
7		?G			
8		?H			
9		!I			
10		?J			
11		K			
12		?OTHERWISE			
13		L			

A default behaviour tree is, like the OTHERWISE statement, executed when an event is received that does not match any of the events being evaluated. However, the default behaviour is specified once in the test case and is then present for all receive events or sets of alternatives. This is shown in Example_5.

Specifically, this means that when a receive statement is met, a snapshot of the queues is taken, and if there is one, the event at the head of the queue is evaluated against the receive constraint and any other alternatives at the same level of indentation. If none are matched, then the event is also evaluated against any alternatives provided in the default. If the event does not match any of the default alternatives (which can include an OTHERWISE), then the event remains in the queue and the evaluation loop is repeated. If the event does match an alternative in the default, then execution continues following the default behaviour (!Y in this example). One very useful feature of defaults is that they can contain a RETURN statement. When this is reached, it returns control back to the original evaluation loop that was running when the default was entered, only now the message that was processed in the default has been removed from the queue. This allows the processing of messages that are unexpected, but not illegal, such as measurement reports, which can be sent routinely by a

Test case dynamic behaviour
Test case name : Example_5
Defaults: Def1

No.	Label	Behaviour description	Constraint ref	Verdict	Comments
1		!A			
2		?B			
3		!C			
4		?D			
5		?E			
Def1					
1		?X			
2		!Y			

which expands to

Test case dynamic behaviour
Test case name : Example_5_expanded
Defaults : Def1

No.	Label	Behaviour description	Constraint ref	Verdict	Comments
1		!A			
2		?B			
3		!C			
4		?D			
5		?E			
6		?X			
7		!Y			
8		?X			
9		!Y			

UE, but may be difficult to plan into the flow of a test. A default terminated by a RETURN allows these messages to be processed silently without affecting the test itself.

A.1.9 Test Steps

Like conventional languages, TTCN also provides a mechanism equivalent to a subroutine, where a behaviour tree can be defined that can be called from other trees. The process of calling a test step is called 'attachment', and it is denoted by a '+' character followed by the name of the test step to be executed. This is shown in the following example.

Test steps can be parameterized and the parameters given when the test step is attached, and they can also be nested, in that a test step can attach to (or call) another test step. When a execution of a test step reaches the end of a branch, it implicitly returns back to the line after and at the next level of indentation in the tree it was called from, if one exists, otherwise

TEST_CASE1
A
?B
+STEP1
D

STEP1
X
?Y
Z

TEST_CASE1
A
?B
X
?Y
Z
D

(i) TEST_CASE1 and the test step STEP1 it attaches

(ii) TEST_CASE1 expanded to show actual execution flow

execution stops. Test steps are usually declared globally and can be called from any test case, or other test step, within a test suite. However, the exception to this is a type of test step called a local tree. Local trees are defined within a test case, and can only be called from that test case, otherwise they operate in the same way. They can be very useful for improving structure and readability.

A.1.10 Program Flow Control

Three constructs are provided to allow the execution flow to be redirected: REPEAT ... UNTIL, GOTO and qualifiers. A qualifier is a Boolean expression placed inside brackets:

> [*Boolean expression*]

If the expression evaluates to TRUE then execution continues with the statements following the qualifier, or if it is the only statement on the line, the next line if it has a higher level of indentation. It is equivalent to an 'if' statement in other languages.

The REPEAT statement has the following syntax:

> REPEAT *test step identifier* UNTIL *qualifier*

The test step is executed and then the qualifier is evaluated. If the qualifier evaluates FALSE, then the test step is repeated and the qualifier re-evaluated. If it evaluates TRUE, then execution continues and the next statement is executed providing it has a higher level of indentation.

Finally, GOTO is followed by a label identifier. This relates to the label column in the dynamic behaviour tables used for earlier examples. Execution will continue from the line which contains the label identifier in the label column.

A.1.11 TTCN and ASN.1

Most things in TTCN are defined using tables; this being the 'tabular notation', but ASN.1 has become a popular way of defining message syntax, especially in communications protocols. TTCN allows the use of ASN.1 as an alternative to tables for defining simple and structured types, PDUs and ASPs. This is particularly relevant for UMTS, as ASN.1 is used to define the RRC formal message syntax. The 3GPP conformance test suites import this ASN.1

directly from the RRC specification TS 25.331. This ensures consistency between the test cases and the core specifications and reduces the chance for errors.

A.1.12 Test Suite Structure

A test suite is grouped into four sections:

Section	Purpose
Test suite overview	This is just descriptive text and has no role in the execution of the tests. It contains mainly indexes and lists of many of the tables within the test suite and in particular has a list of the test cases and test steps included.
The declarations part	In this part all of the objects used in the test suite are declared. This includes • User-defined simple type definitions • Structured type definitions • Test suite (global) parameters and constants • Test case selection expressions • PCOs and PCO types • Timers • ASPs • PDUs.
The constraints part	In this part, all of the constraints are declared and values that are not set from parameters are provided.
The dynamic part	This contains the behaviour trees for all of the test cases and test steps that make up the suite. This also includes the behaviour trees that make up the default behaviours.

A.1.13 Approach to Testing

This appendix is only intended to give a brief overview of the TTCN 2 language. For a test engineer intending to use the language in practice, detailed training courses are run by many of the TTCN tool vendors. Like any software, TTCN code needs to be architected and well structured if it is to be maintained in the future. This requires a disciplined approach through the complete life cycle of the development and upkeep of the ATSs. In the case of the 3GPP test suites, considerable thought was put into both the structure and the architecture of each ATS. In particular, the structure reflects the needs of a project of this scale to allow work on different parts of an ATS to proceed in parallel, without creating conflicts with engineers trying to modify the same code. In TTCN, this is especially important as merging of TTCN is very difficult. However, probably the most important part of the 3GPP process was introduced in 2002, when a regular 'build-and-test' cycle was put in place. Every few weeks, the changes and fixes from the individual ETSI task force engineers working on the TTCN are brought together into a single release. This is released to the TTCN verification teams. These are engineers working mainly in the industry, who build the TTCN into executable suites for their test platforms and run all the test cases. A regression report is sent back to ETSI task force and, if needed, fixes are applied and the TTCN re-released. This process has enabled the TTCN to move forwards at a considerable pace while keeping a high level of stability and quality. The process developed by 3GPP and ETSI serves as an excellent reference for any large-scale TTCN test suite development.

Glossary

I. Common Abbreviations

3GPP	3rd-generation partnership project
ACK	Acknowledge
ACLR	Adjacent Channel Leakage Power Ratio
A-GPS	Assisted GPS
AMR	Adaptive Multirate (speech codec)
AM-RLC	Acknowledged Mode RLC
AMR-WB	AMR Wideband
APN	Access Point Name
ARIB	Association of Radio Industries and Businesses
ARQ	Automatic Repeat reQuest
AS	Access Stratum
ASIC	Application-Specific Integrated Circuit
ASN.1	Abstract Syntax Notation One
ASP	Abstract Service Primitive
ATIS	Alliance for Telecommunications Industry Solutions
ATS	Abstract Test Suite
AuC	Authentication Centre
AWGN	Additive White Gaussian Noise
BER	Bit Error Rate
BLER	Block Error Rate or Block Error Ratio
BMC	Broadcast/Multicast Control
BSC	Base Station Controller
BTS	Base Transceiver Station
CCSA	China Communication Standards Association
CDMA	Code Division Multiple Access
CN	Core Network
Codec	Coder/Decoder
C-RNTI	Cell Radio Network Temporary Identifier
CS	Circuit Switched

Testing UMTS: Assuring Conformance and Quality of UMTS User Equipment   Dan Fox
© 2008 John Wiley & Sons, Ltd

dBc	Decibels Relative to the Carrier
dBm	Decibels Relative to 1 Milliwatt (mW)
DECT	Digital Enhanced Cordless Telephony
DNS	Domain Name Service
DRX	Discontinuous Reception
DTX	Discontinuous Transmission
E_c	Energy per Chip
E_c/N_0	Ratio of the energy per chip to the noise spectral density
EDGE	Enhanced Data rates for GSM Evolution
EGPRS	Enhanced GPRS
EMC	Electromagnetic radiation Compliance
ETSI	European Telecommunications Standards Institute
E-UL	Enhanced Uplink
EVM	Error Vector Magnitude
FDD	Frequency Division Duplex
FDMA	Frequency Division Multiple Access
FPGA	Field Programmable Gate Array
FRC	Fixed Reference Channel
GCF	Global Certification Forum
GERAN	GSM/EDGE Radio Access Network
GGSN	Gateway GPRS Support Node
GMSC	Gateway MSC
GMSK	Gaussian Minimum Shift Keying
GPRS	General Packet Radio System
GPS	Global Positioning System
GRR	GPRS Radio Resource
GSM	Groupe System Mobile (*Fr.*) (also known as Global System Mobile)
HARQ	Hybrid ARQ
HHO	Hard Handover
HLR	Home Location Register
HSDPA	High-Speed Downlink Packet Access
HSPA	High-Speed Packet Access
HSUPA	High-Speed Uplink Packet Access (see also E-UL)
ICS	Implementation Conformance Statement (see also PICS)
IETF	Internet Engineering Task Force
IMS	Integrated Multimedia Subsystem
IMSI	International Mobile Subscriber Identity
I_{oc}	The power spectral density (integrated in a noise bandwidth equal to the chip rate and normalized to the chip rate) of a band limited white noise source (simulating interference from cells which are not defined in a test procedure) as measured at the UE antenna connector
I_{or}	The total transmit power spectral density (integrated in a bandwidth of $(1 + a)$ times the chip rate and normalized to the chip rate) of the downlink signal at the Node B antenna connector, where a is the roll-off factor of the root-raised cosine filter used to shape the transmitted signal
IP	Internet Protocol
ITU	International Telecommunications Union
I_{ub}	Interface between the Node B and the RNC
I_{ur}	Interface between RNCs
LLC	Logical Link Control
LTE	Long-Term Evolution
MAC	Medium Access Control
MBMS	Multimedia Broadcast/Multicast Service
MIMO	Multiple Input Multiple Output
MMS	Multimedia Messaging Service
MSC	Mobile Switching Centre
NACK	Negative acknowledge
NAS	Nonaccess Stratum

Node B	The node within the network responsible for radio transmission and reception to/from the UE
OCNS	Orthogonal Channel Noise Simulation
OFDMA	Orthogonal FDMA
OVSF	Orthogonal Variable Spreading Factor
PAR	Peak to Average Power Ratio (sometimes abbreviated to PAPR)
PCO	Point of Control and Observation
PCU	Packet Control Unit
PD	Protocol Discriminator
PDA	Personal Digital Assistant
PDCP	Packet Data Convergence Protocol
PDP	Packet Data Protocol
PDU	Protocol Data Unit
PhCH	Physical Channel
PHY	Physical Layer
PICS	Protocol Implementation Conformance Statement
PIXIT	Protocol Implementation eXtra Information for Testing
PLMN	Public Land Mobile Network
PS	Packet Switched
PTCRB	PCS Type Certification Review Board
P-TMSI	Packet Temporary Mobile Subscriber Identity
QAM	Quadrature Amplitude Modulation
QoS	Quality of Service
QPSK	Quadrature Phase Shift Keying
R&TTE	Radio and Telecommunications Terminal Equipment
RAB	Radio Access Bearer
RAN	Radio Access Network
RAT	Radio Access Technology
RB	Radio Bearer
RFC	Request For Comments
RLC	Radio Link Control
RM	Rate Matching
RMC	Reference Measurement Channel
RNC	Radio Network Controller
RNTI	Radio Network Temporary Identity
RR	Radio Resource
RRC	Radio Resource Control
RRM	Radio Resource Management
RSCP	Received Signal Code Power
RSSI	Received Signal Strength Indicator
RTP	Real-time Transport Protocol
SAP	Service Access Point
SAPI	Service Access Point Identifier
SAR	Specific Absorption Ratio
SDU	Service Data Unit
SEM	Spectrum Emission Mask
SF	Spreading Factor
SFN	System Frame Number
SGSN	Serving GPRS Support Node
SHO	Soft Handover
SIB	System Information Block
SIM	Subscriber Interface Module
SIR	Signal to Interference Ratio
SMG	Special Mobile Group
SMS-SC	Short Message Service Switching Centre
SNR	Signal to Noise Ratio
SRB	Signalling Radio Bearer

SRNS	Serving Radio Network Subsystem
SS	System Simulator
TB	Transport Block
TBS	Transport Block Set
TCP	Transmission Control Protocol
TDD	Time Division Duplex
TDMA	Time Division Multiple Access
TD-SCDMA	Time Division-Synchronous Code Division Multiple Access
TFCI	Transport Format Combination Indicator
TFCS	Transport Format Combination Set
TFS	Transport Format Set
TFT	Traffic Flow Template
TI	Transaction Identifier
TM-RLC	Transparent Mode RLC
TMSI	Temporary Mobile Subscriber Identity
TPC	Transmitter Power Control
TrCH	Transport Channel
TSG	Technical Specification Group
TTC	Telecommunication Technology Committee
TTCN 2	Tree and Tabular Combined Notation (Version 2)
TTCN 3	Test and Testing Control Notation (Version 3)
TTI	Transmission Time Interval
UARFCN	UTRA Absolute Radio Frequency Channel Number
UDP	User Datagram Protocol
UE	User Equipment
UI	User Interface
UICC	Universal Integrated Circuit Card
UM-RLC	Unacknowledged Mode RLC
UMTS	Universal Mobile Telecommunications System
URA	UTRAN Registration Area
URL	Uniform Resource Locator
U-RNTI	UTRAN Radio Network Temporary Identity
USAT	USIM Applications Toolkit
USB	Universal Serial Bus
USIM	UMTS SIM
UTRA	UMTS Terrestrial Radio Access
UTRAN	UMTS Terrestrial Radio Access Network
VLR	Visitor Location Register
VoIP	Voice over IP
WCDMA	Wideband CDMA
XML	eXtensible Mark-up Language

II. Common Physical, Transport and Logical Channel Names

S-CCPCH	Secondary Common Control Physical Channel
PRACH	Physical Random Access Channel
P-CCPCH	Primary Common Control Physical Channel
CPICH	Common Pilot Channel
P-CPICH	Primary Common Pilot Channel
S-CPICH	Secondary Common Pilot Channel
SCH	Synchronization Channel
P-SCH	Primary Synchronization Channel
S-SCH	Secondary Synchronization Channel

PICH	Paging Indicator Channel
AICH	Acquisition Indicator Channel
DPCH	Dedicated Physical Channel
DPDCH	Dedicated Physical Data Channel
DPCCH	Dedicated Physical Control Channel
E-DPDCH	Enhanced Dedicated Physical Data Channel
E-DPCCH	Enhanced Dedicated Physical Control Channel
HS-DPCCH	High-Speed Downlink Physical Control Channel
HS-SCCH	High-Speed Shared Control Channel
RACH	Random Access Channel
BCH	Broadcast Channel
PCH	Paging Channel
FACH	Forward Access Channel
DCH	Dedicated Channel
E-DCH	Enhanced Dedicated Channel
BCCH	Broadcast Control Channel
PCCH	Paging Control Channel
CCCH	Common Control Channel
DCCH	Dedicated Control Channel
DTCH	Dedicated Traffic Channel
CTCH	Common Traffic Channel

References

References to figures and tables used within the book.

Information from the following 3GPP specification figures and tables has been used as a reference for various figures and tables within the book. In general, the information has been simplified or reformatted to draw out the issues related to testing, but the original references are provided here.

Figures and tables have been reproduced from the 3GPP specifications with the kind permission of the ETSI Director General.

3GPP™ TSs and TRs are the property of ARIB, ATIS, ETSI, CCSA, TTA, and TTC who jointly own the copyright for them. They are subject to further modifications and are therefore provided to you "as is" for information purposes only. Further use is strictly prohibited.

Figure 8.3	Source: TS 25.213 v7.1.0, Figure 4
Figure 8.5	Source: TS 25.211 v7.1.0, Figure 1
Table 8.2	Source: TS 34.108 v7.0.0, table in clause 6.10.2.4.1.32.2.1.1
Figure 8.8	Source: TS 25.213 v7.1.0, Figure 1A
Figure 8.9	Source: TS 25.101 v7.7.0, Figure A.1
Table 8.4	Source: TS 25.101 v7.7.0, Tables 6.1 and 8.5
Table 8.6	Source: TS 25.101 v7.7.0, Table 6.4
Table 8.7	Source TS 34.121-1 v7.4.0, Table 5.4.1.3
Figure 8.16	Source: TS 25.101 v7.7.0, Figure 6.2
Figure 8.17	Source: TS 34.121-1 v7.4.0, Figure 5.6.1
Figure 8.18	Source: TS 25.101 v7.7.0, Figure 6.1
Table 8.8	Source: TS 25.101 v7.7.0, Table 6.10
Table 8.9	Source: TS 25.101 v7.7.0, Tables 6.12 and 6.13
Table 8.10	Source: TS 25.101 v7.7.0, Table 6.14
Table 8.11	Source: TS 25.101 v7.7.0, Tables C.6 and C.7
Table 8.12	Source: TS 25.101 v7.7.0, Table 7.9A
Table 8.13	Source: TS 25.101 v7.7.0, Table 7.10
Table 8.14	Source: TS 25.101 v7.7.0, Tables 8.7 and 8.8
Figure 9.1	Source: TS 25.301 v7.1.0, Figure 4
Figure 9.2	Source: TS 25.321 v7.4.0, Figures 4.2.2.1 and 4.2.3.1
Figure 9.4	Source: TS 25.321 v7.4.0, Figure 4.2.3.1.1

Testing UMTS: Assuring Conformance and Quality of UMTS User Equipment   Dan Fox
© 2008 John Wiley & Sons, Ltd

Selection of Useful 3GPP Specifications

UTRAN and UMTS core network signaling

TS 23.122   Nonaccess stratum (NAS) functions related to mobile station (MS) in idle mode
TS 24.008   Mobile radio interface layer 3 specification, core network protocols, stage 3
TS 24.011   Point-to-point (PP) Short Message Service (SMS) support on mobile radio interface
TS 25.101   User equipment (UE) radio transmission and reception (FDD)
TS 25.133   Requirements for support of radio resource management (FDD)
TS 25.171   Requirements for support of Assisted Global Positioning System (A-GPS), frequency division duplex (FDD)
TS 25.201   Physical layer – general description
TS 25.211   Physical channels and mapping of transport channels onto physical channels (FDD)
TS 25.212   Multiplexing and channel coding (FDD)
TS 25.213   Spreading and modulation (FDD)
TS 25.214   Physical layer procedures (FDD)
TS 25.215   Physical layer, measurements (FDD)
TS 25.301   Radio interface protocol architecture
TS 25.302   Services provided by the physical layer
TS 25.303   Interlayer procedures in connected mode
TS 25.304   UE procedures in idle mode and procedures for cell reselection in connected mode
TS 25.306   Radio access capabilities
TS 25.306   UE radio access capabilities
TS 25.307   Requirements on UEs supporting a release-independent frequency band
TS 25.308   High-speed downlink packet access (HSDPA), overall description, stage 2
TS 25.309   FDD enhanced uplink, overall description, stage 2
TS 25.321   Medium access control (MAC) protocol specification
TS 25.322   Radio link control (RLC) protocol specification
TS 25.323   Packet data convergence protocol (PDCP) specification
TS 25.324   Broadcast/multicast control (BMC)
TS 25.331   Radio resource control (RRC), protocol specification
TS 26.071   AMR speech codec, general description
TS 26.171   Speech codec speech processing functions, adaptive multi-rate – wideband (AMR-WB) speech codec, general description

USIM specifications

TS 31.101	UICC-terminal interface, physical and logical characteristics
TS 31.111	Universal Subscriber Identity Module (USIM) Application Toolkit (USAT)
TS 31.120	UICC-terminal interface, physical, electrical, and logical test specification
TS 31.121	UICC-terminal interface, USIM application test specification
TS 31.122	USIM conformance test specification
TS 31.124	Mobile equipment (ME) conformance test specification, USAT conformance test specification

GSM specifications

TS 23.060	General Packet Radio Service (GPRS), service description, stage 2
TS 43.051	GSM/EDGE Radio Access Network (GERAN) overall description, stage 2
TS 43.055	Dual transfer mode (DTM), stage 2
TS 43.064	GPRS, overall description of the GPRS radio interface, stage 2
TS 44.001	Mobile station-base station system (MS-BSS) interface general aspects and principles
TS 44.003	MS-BSS interface channel structures and access capabilities
TS 44.004	Layer 1, general requirements
TS 44.013	Performance requirements on mobile radio interface
TS 44.014	Individual equipment-type requirements and interworking, special conformance testing functions
TS 44.018	Mobile radio interface layer 3 specification, RRC protocol
TS 44.060	GPRS, MS-BSS interface, radio link control/medium access control (RLC/MAC) protocol
TS 44.064	Mobile station-serving GPRS support node (MS-SGSN), logical link control (LLC) layer specification
TS 44.065	MS-SGSN, subnetwork-dependent convergence protocol (SNDCP)
TS 45.001	Physical layer on the radio path, general description
TS 45.002	Multiplexing and multiple access on the radio path
TS 45.003	Channel coding
TS 45.004	Modulation
TS 45.005	Radio transmission and reception
TS 45.008	Radio subsystem link control

Test-related specifications

TS 26.074	AMR speech codec, test sequences
TS 26.132	Speech and video telephony terminal acoustic test specification
TS 26.174	Speech codec speech processing functions, AMR-WB speech codec test sequences
TS 34.108	Common test environments for UE, conformance testing
TS 34.109	Terminal logical test interface, special conformance testing functions
TS 34.121-1	UE conformance specification, radio transmission and reception (FDD), Part 1: Conformance specification
TS 34.121-2	UE conformance specification, radio transmission and reception (FDD), Part 2: Implementation conformance statement (ICS)
TS 34.123-1	UE conformance specification, Part 1: Protocol conformance specification
TS 34.123-2	UE conformance specification, Part 2: ICS specification
TS 34.123-3	UE conformance specification, Part 3: Abstract test suites (ATSs)
TS 34.124	Electromagnetic compatibility (EMC) requirements for mobile terminals and ancillary equipment
TS 34.171	Terminal conformance specification, A-GPS, FDD
TS 51.010-1	MS conformance specification, Part 1: Conformance specification

TS 51.010-2	MS conformance specification, Part 2: Protocol implementation conformance statement (PICS) proforma specification
TS 51.010-3	MS conformance specification, Part 3: Layer3 (L3) ATS
TS 51.010-5	MS conformance specification, Part 5: Inter-radio access technology (RAT) (GERAN/UTRAN) interaction ATS

TTCN specifications

ISO/IEC 9646-1	Information technology – Open systems interconnection – Conformance testing methodology and framework – Part 1: General concepts
ISO/IEC 9646-2	Information technology – Open systems interconnection – Conformance testing methodology and framework – Part 2: Abstract test suite specification
ISO/IEC 9646-3	Information technology – Open systems interconnection – Conformance testing methodology and framework – Part 3: The tree and tabular combined notation (TTCN)
ISO/IEC 9646-4	1994, Information technology – Open Systems Interconnection – Conformance testing methodology and framework – Part 4: Test realization
ISO/IEC 9646-5	Information technology – Open systems interconnection – Conformance testing methodology and framework – Part 5: Requirements on test laboratories and clients for the conformance assessment process

General Interest

TR 21.900	Technical specification group working methods
TR 21.905	Vocabulary for 3GPP specifications
TS 23.002	Network architecture
TS 23.101	General UMTS architecture
TS 23.107	Quality-of-service (QoS) concept and architecture
TS 23.207	End-to-end QoS concept and architecture
TS 23.110	UMTS access stratum services and functions
TS 23.228	IP multimedia subsystem (IMS), stage 2
TS 25.401	UTRAN overall description
TS 27.007	AT command set for UE

Further reference for some areas not covered in this book

TS 25.346	Introduction of the Multimedia Broadcast/Multicast Service (MBMS) in the Radio Access Network (RAN), stage 2
TS 23.246	MBMS, architecture and functional description
TS 23.271	Functional stage 2 description of location services (LCS)
TS 24.010	Mobile radio interface layer 3 – Supplementary services specification – General aspects
TS 25.141	BS conformance testing (FDD)
TS 33.102	3G security, security architecture
TS 34.122	Terminal conformance specification, radio transmission and reception (TDD)
TS 34.229-1	Internet protocol (IP) multimedia call control protocol based on session initiation protocol (SIP) and session description protocol (SDP), Part 1: Protocol conformance specification
TS 34.229-2	IP multimedia call control protocol based on SIP and SDP, Part 2: Implementation Conformance statement
TS 34.229-3	IP multimedia call control protocol based on SIP and SDP, Part 3: Abstract test suite (ATS)

TR 34.902	Derivation of test tolerances for multi-cell radio resource model (RRM) conformance tests
TR 34.926	Electromagnetic compatibility (EMC), Table of international requirements for mobile terminals and ancillary equipment
TR 43.058	Characterization, test methods, and quality assessment for handsfree MSs

Books

Holma, H. and Toskala, A. (ed.) (2004) *WCDMA for UMTS: Radio Access for Third Generation Mobile Communications*, 3rd edn. John Wiley & Sons.

Kreher, R. and Ruedebusch, T. (2007) *UMTS Signaling: UMTS Interfaces, Protocols, Message Flows and Procedures Analyzed and Explained*, 2nd edn. John Wiley & Sons.

Kaaranen, H., Ahtiainen, A., Laitinen, L., Naghian, S., Niemi, V., and Naghian, S. (2001) *UMTS Networks: Architecture, Mobility and Services*, John Wiley & Sons.

Index

3GPP (3rd Generation Partnership Project) 4–5, 8, 10, 12–13, 20, 30–5, 37, 39, 40, 48, 50, 51–6, 59, 61, 72, 78, 80, 82, 86, 89, 90, 108, 163, 182, 207, 216, 218, 219–20, 222, 225, 226, 240–1

Abstract Service Primitive (ASP) 233–5
Abstract Syntax Notation One (ASN.1) 156, 220, 231–2, 239
Abstract Test Suite (ATS) 224, 239–40
Acceptable Cell 153, 198–9
Acceptance 6, 7, 20, 49, 53, 54, 59, 64–5, 67–9, 171, 197
Access Point Name (APN) 173, 193
Access Service Class (ASC) 125, 126, 131–2
  ASCII 47, 231
Access Stratum (AS) 40, 41, 47, 217
Acknowledged Mode (AM) RLC 132, 135, 137–41, 144, 186, 208
Acquisition Indicator Channel (AICH) 83, 85, 98, 131, 132, 185
Activate PDP context 193–4
Activate secondary PDP context 174
Active set 161–2, 166, 195–6, 204–6
Adaptive Modulation and Coding (AMC) 207
Adaptive modulation 82, 89, 92, 105–8, 112, 119, 149, 207–8, 222
Adaptive Multi-Rate Wideband (AMR-WB) 35, 52
Adaptive Multi-Rate (AMR) speech codec 35, 52, 189
Additive White Gaussian Noise (AWGN) 116
Adjacent Channel Leakage Power Ratio (ACLR) 73, 102, 103, 104, 212
Adjacent Channel Selectivity 110–1
Adversarial testing 26

ALOHA 130–1
AM-RLC, *see* Acknowledged Mode (AM) RLC
AS, *see* Access Stratum (AS)
Assisted Global Positioning System (A-GPS) 35, 41, 42, 50, 51, 56, 59
Association of Radio Industries and Business (ARIB) 11, 12, 52, 59
AT commands 27, 189
Attach 170–1, 174, 188–9, 191–3, 239
Authentication Centre (AuC) 147, 148, 169
Authentication 41, 95, 109, 151, 169–70, 188, 190–4
Authentication vector 169, 191
Automatic Repeat Request (ARQ) 135–6
Automation 27, 221

Barred or Reserved Cell 67, 198, 201
Base constraint 230
Base Station Controller (BSC) 149
Base Transceiver Station (BTS) 149
Baseband 24, 37, 44, 93, 105, 116, 118–19, 183, 222
Birth-death 116
Bit Error Rate (BER) 106, 117
Block Error Rate/Block Error Ratio (BLER) 45, 59, 116–17, 157, 166, 203, 205
Blocking characteristics 111–12
Broadcast Channel (BCH) 122, 123, 124, 125, 150, 155, 185, 186, 197
  see also Broadcast Control Channel (BCCH)
Broadcast Control Channel (BCCH) 83, 122–6, 128, 133, 185–6
Broadcast Control Functional Entity (BCFE) 150
Broadcast/Multicast Control (BMC) 48, 123, 146, 177

Testing UMTS: Assuring Conformance and Quality of UMTS User Equipment   Dan Fox
© 2008 John Wiley & Sons, Ltd